Engineering a High-Tech Business

Entrepreneurial Experiences and Insights

Engineering a High-Tech Business

Entrepreneurial Experiences and Insights

José Miguel López-Higuera
Brian Culshaw
Editors

Bellingham, Washington USA

Library of Congress Cataloging-in-Publication Data

Engineering a high-tech business: entrepreneurial experiences and insights / Jose López-Higuera and Brian Culshaw, Editors.
p. cm. -- (Press monograph ; pm 182)
ISBN 978-0-8194-7180-2
1. Industrial engineering. I. López-Higuera, José Miguel. II. Culshaw, B.

T56.E58 2008
620.0023--dc22

2007050209

Published by

SPIE
P.O. Box 10
Bellingham, Washington 98227-0010 USA
Phone: +1 360.676.3290
Fax: +1 360.647.1445
Email: Books@SPIE.org
spie.org

Printed in the United States of America.

This book is dedicated to the people who work hard every day advancing science and the technology to contribute to improving our society.

To our parents, recognizing all that they did for us.

The Editors

Table of Contents

List of Acronyms ix

Introduction 1

Part I: Reflections, Motives, and Money 5

1. Some Suggestions from an Economist to a New High-Tech Starter *9*
Guillermo de la Dehesa

2. The Academic Entrepreneur: An Oxymoron? *23*
Brian Culshaw

3. Money *29*
Stuart Barnes

4. Confessions of a Start-Up Junkie *35*
Mike Redman

5. Being an "Intrapreneur" and an Entrepreneur in the Optoelectronics Industry *51*
Michael S. Lebby

Part II: Some Case Studies 59

6. Mirada Solutions: The Case Study of a University Spin-Off *61*
Miguel Mulet Parada and Sir Michael Brady

7. Building a Company the Old-Fashioned Way: Meadowlark Optics, Inc. *77*
Tom Baur and Garry Gorsuch

8. Building a Lasting Optical Design and Manufacturing Company *87*
Jay Kumler

9. The Life and Times of a High-Tech Entrepreneur *97*
Colleen Fitzpatrick

10. A Case from Russia: IPG Photonics *107*
Valentin Gapontsev

11. Wacko WYKO *113*
James C. Wyant

12. The Ocean Optics Story in a Nutshell *121*
Mike Morris

13. Experiences Starting a Nano-Sized Company *129*
Richard O. Claus and Jennifer H. Lalli

14. The Story of Fiberonics *137*
Jaspreet Singh

15. Founding a Fiber-Optic Component and Sensor Business *147*
Ingolf Baumann

16. How to Start a Small High-Tech Business in Troutdale, Oregon *155*
Eric Udd

17. SMARTEC: Bringing Fiber-Optic Sensors into Concrete Applications *161*
Daniele Inaudi and Nicoletta Casanova

18. "An Earth Odyssey" or Fibersensing *169*
Alberto Maia, Francisco Araújo, José Luís Santos, Luís Ferreira, and Pedro Alves

19. The First Years of Crystal Fibre A/S from a University Perspective *181*
Anders Bjarklev and Jes Broeng

20. Multiwave Photonics: Building a Fiber Optics Company in Portugal *189*
Jose R. Salcedo

Part III: Supporting the Entrepreneur 193

21. Developing High-Tech Companies in Spain and Portugal *195*
Javier Ulecia

22. Intellectual Property in High-Tech Entrepreneurship *205*
Joseph E. Gortych

23. Support for a Young Company *221*
Chris Gracie

Part IV: The Universities 231

24. Strategic Support: The Case of the Technical University of Madrid *233*
Gonzalo León Serrano

25. University Research and the Optics Industry *243*
Faramarz Farahi

Postscript: Some Concluding Thoughts 251

Appendix: Due Diligence Check List 253

Editor and Author Biographies 261

List of Acronyms

ACS	Advanced Cameras for Surveys
AdI	Portuguese Innovation Agency
AMP	A connectors company
AOS	Advanced Optics Solutions
AT&T	American Telephone and Telegraph
C Corporation	Standard Corporation, an entity apart from its owners that may engage in business, issue contracts, sue, be sued, and pay taxes
CAD	Computer Aided Design
CAGR	Compounded Annual Growth Rate
CEO	Chief Executive Officer
CMP	Chemical Mechanical Planarization
CNC	Computer Numerical Control
CT	Chamber of Trade
CT	Computed Tomography
CTC	Chamber of Trade and Commerce
CTI	The company that invented PET
CTO	Chief Technical Officer
DARPA	Defense Advanced Research Projects Agency
DLP	Digital Light Processing
DMD	Digital Micromirror Device
DOC	Digital Optics Corporation
DoD	U.S. Department of Defense
DOE	Department of Energy
DTI	Department of Trade and Industry
DTU	Technical University of Denmark
DUV	Division for Utilities and Ventures
DWDM	Dense Wavelength Division Multiplexing
EBITDA	Earnings Before Interest, Taxes, Depreciation, and Amortization
EEC	European Economic Community
EFDA	Erbium-Doped Fiber Amplifier
EPA	Environmental Protection Agency
EPFL	Swiss Federal Institute of Technology of Lausanne
ERP	Enterprise Resource Planning
ESA	Electrostatic Self-Assembly
FBG	Fiber Bragg Gratings
FDA	U.S. Food and Drug Administration
FDI	Foreign Direct Investment

List of Acronyms (cont.)

FEORC	Fiber and Electro-Optics Research Center at Virginia Tech
FOCAS	Fiber Optic Cables Accessories and Sensors
FSCR	Fundo de SindiÇao de Capital de Risco
IAPMEI	Institute for the Support of Small and Medium-Sized Enterprises
ICOIA	International Coalition of Optoelectronics Association
IDC	Global provider of market intelligence, advisory services, and events for information technology, telecommunications, and consumer technology markets
IEE	Institution of Electrical Engineers
IEEE	Institute of Electrical and Electronic Engineers
IIT	Indian Institute of Technology
ILI	Industrial Laser Institute
IMM	Institute of Material Mechanics
IP	Intellectual Property
IPO	Initial Public Offering
IPR	Intellectual Property Rights
ISO	International Organization for Standardization
ITT	International Telephone and Telegraph Corporation
LDA	Local Development Agency
LED	Light Emitting Diode
LEOS	Lasers and Electro-Optics Society
MBA	Master in Business Administration
MD	Managing Director
MEMS	Micro-Electro-Mechanical System
MIT	Massachusetts Institute of Technology
MR	Magnetic Resonance
MVL	Medical Vision Laboratory
NASA	National Aeronautics and Space Administration
NDA	Nondisclosure Agreement
NEST	Novas Empresas de Suporte Tecnológico
NICMOS	Near Infrared Camera and Multi-Object Spectrograph
NIH	National Institutes of Health
NLRB	National Labor Relations Board
NMR	Nuclear Magnetic Resonance
NPDU	New-Product-Development Unit
NSF	National Science Foundation
NSGT	Next-Generation Space Telescope
NTB	New Technology-Based Company
OCT	Optical Coherent Tomography
OEM	Original Equipment Manufacturer
OIDA	Optoelectronics Industry Development Association

List of Acronyms (cont.)

OITDA	Optoelectronics Industry and Technology Development Association
OSA	Optical Society of America
OSHA	Occupational Safety and Health Administration
OTRI	Offices for Technology Transfer
P&L	Profit and Loss Statement
PABX	Private Automatic Branch Exchange
PCB	Printed Circuit Board
PCT	Patent Cooperation Treaty
PDA	Personal Digital Assistant
PDMS	Polydimethylsiloxane
PET	Positron Emission Tomography
PRIME	Incentives program for the modernization of economic activities
PTAP	Photonics Technology Access Program
PTC	Product Technology Center
QC	Quality Central, Quality Control
R&D	Research and Development
RAS	Russian Academy of Science
RSNA	Radiological Society of North America
S Corporation	Standard Corporation that has elected a special tax status with the Internal Revenue Service
S&T	Science and Technology
SAS	Soviet Academy of Sciences
SBIR	Small Business Innovative Research
SDA	Scottish Development Agency
SEM	Scanning Electron Microscope
SID	The Society for Information Display
SIM	Space Interferometric Mission
SIRTF	Space Infrared Telescope Facility
SME	Small to Medium Enterprise
SME	Subject-Matter Expert
SOA	Scottish Optoelectronic Association
SOFIA	Stratospheric Observatory for Infrared Astronomy
SOFO	Surveillance des Ouvrages par Fibres Optiques or Low-Coherence Interferometric Displacement Sensor System
SPECT	Single Photon Emission Computed Tomography
SPIE	Society of Photo-Optical Instrumentation Engineers
STC	Standard Telephones and Cables
STIS	Space Telescope Imaging Spectrograph
SUNX	A sensors company
TFT	Thin-Film Transistor

List of Acronyms (cont.)

TTOM	Technology Transfer in Optoelectronics and Microelectronics
UHF	Ultra High Frequency
UNCC	University of North Carolina, Charlotte
UPM	Technical University of Madrid
URL	Uniform Resource Locator
USPTO	United States Patent and Trademark Office
VAT	Value-Added Tax
VC	Venture Capital
VP	Vice President
VPS&M	Vice President Sales and Marketing
YDFA	Ytterbium-Doped Fiber Amplifier

Introduction

This is a book of stories, perhaps even fables, of the technological entrepreneur.

We put this book together for the science and technology community. Our principal objective has been to convey the adventure and the spirit of the entrepreneurial engineer rather than present recipes and approaches. We feel that this perspective becomes increasingly important as young scientists and engineers emerge into a business environment that is more and more the domain of the small energetic enterprise. Indeed, many will argue that most current technological innovation emanates from such activity. The technical excitement, the control of one's own destiny, the apparent independence from corporate politics, and the prospects of accumulating even a modest sum can all attract the young engineer.

And we now see "entrepreneurship" appearing in university education. Many see this as an immense benefit, but we would argue that entrepreneurship education can be at best an oxymoron and even counterproductive. It is this uneasy prospect that entrepreneurial skills appear to be encapsulated within a set routine, with a simple, logical, single-valued solution. Yes, there are certainly some common elements, and these come through in the anecdotes presented in the book. However, these common elements are generic, have multi-valued manifestations, and can be summarized briefly.

This unease becomes yet more pronounced when relating entrepreneurial activity to science and technology. The practitioners in these disciplines instinctively endeavor to formulate a problem in a relatively simplistic, closed fashion and thereafter seek the closed solution, with logical, justifiable steps from beginning to end. The scientist will cling to this as the central tenet of his faith. The engineer may admit to more diversity. There are, after all, several workable solutions to engineering a highway bridge or an electronic circuit, but these remain closed and fit a technical specification. As we advance into design, sociology and economics, even finance and management, formulating the problem becomes increasingly elusive. But solutions, albeit far from closed, have to be found. Entrepreneurship is best regarded as an open solution to an open problem.

The contributions in this short book illustrate the diversity of aims and aspirations, personalities and motivations, necessities and luxuries that comprise just one small sample of the range of technical entrepreneurship. Our hope here is to illustrate that entrepreneurship embraces a huge diversity of people, techniques, and requirements. There is most certainly no single route forward. There is certainly scope for anyone who feels the interest to explore the prospects that this world can offer.

The stereotypes are nothing more than stereotypes. Yes, the brash exhibitionist exists, though probably in the minority. Yes, there are the self-made millionaires, but these are far outnumbered by those achieving many years of comfortable living. Yes, some capitalize on highly protected scientific breakthroughs, but many more engineer a well-known concept into a particular geographical market or social niche.

Our book has been edited into four sections, though we would be the first to admit that the editing is by no means the only solution to presenting the entrepreneurial story. We start by exploring motivation and money through five chapters. Arguably, these are the most basic necessities of entrepreneurial activity, and even they have their diversities, not only at the level of individuals involved but also through societies, cultures, and communities. Indeed, the latter may be more influential. We then move on to 15 short case studies of individual companies whose stories, taken from several countries and cultures, are predominantly in the optoelectronic sector.

Supporting the entrepreneur is our third theme. This has three principal strands. One is the creation of networks and clusters through which the entrepreneurial community can meet with each other and with the world outside. Trade associations, development corporations, and professional societies all have much to contribute. Next is cash. We explore an example from the world of venture capital, though other sources exist, and the case studies highlight numerous examples of what many perceive as more "friendly" funds. The fourth section explores the role of universities in the technology process. As academics, we believe that universities have important, even critical, contributions to make. We also believe that the vast majority of university organizations still have much to learn in handling this process and resolving how best to make their all-important contribution.

The stories, the fables you find in this book most certainly highlight entrepreneurial diversity. There is, though, much in common among all the stories. At the core of each is an enthusiast with a technical idea and a desire to see something happen. This is vital.

Much of the rest is about telling the story, projecting the enthusiasm, maintaining the commitment. Persistence is another essential. The idea you have had and demonstrated in the laboratory, possibly even written a technical paper about, is nothing more than that. The difference between idea and product is enormous. The first basic question is whether anybody cares about the prospects the idea may offer. If they do, then perhaps there is a market; but realizing that market requires a product (lots more work than the idea) and an infrastructure with which to market the product. This infrastructure will, as we all learn, dominate the process.

Changing your idea from the lab into a product and providing the infrastructure to market it takes money. Without both product and marketing, no one ever buys anything. In the end, your company has to pay its bills by selling its wares. This is all very obvious but is often initially surprisingly low on the aspiring entrepreneur's agenda. It often takes some persuasion for the technical initiator to accept that interdisciplinary skills are necessary. These skills extend far beyond the technical diversity required to realize the product and move into the "softer" regimes of marketing, leadership, economics, finance, and management.

There is a need, too, for partnership, whether financial, technical, marketing, applications engineering, trade shows and exhibits, political lobbying, etc. The list goes on, and both the problem statement and solution become increasingly open.

In parallel, the art of telling the story to all in this diverse community becomes increasingly important. Storytelling is a two-way process, and the entrepreneur must dedicate at least as much to listening, absorbing, and interpreting as to conveying his own message. It is almost always those outside the entrepreneur's immediate community who finally support the ideas, whether as investors, endorsers, or customers. Listening and responding to these outsiders is critical to eventual success.

So our short volume aims to give at least some insight into this fascinating, multidimensional, multidisciplinary activity called technological entrepreneurship. We believe it extends far beyond the confines of a self-contained, easily encapsulated, concisely summarized endeavor. It is organic, continuously growing and evolving, and advancing into new domains. It is something that must be lived rather than described, and our contributors have all lived their roles in their particular corner of the entrepreneurial life and all that that implies. Embryonic entrepreneurs who read this book will find that their own perspective will be entirely different, but we hope we shall inspire the confidence, the curiosity, and the initiative to give it a try.

And we have many people to thank, most notably all the contributors who have taken the time from being entrepreneurs to describing their perspective on entrepreneurship. We greatly value their input and insights, and we hope their collective leadership will help inspire others with the essential confidence and commitment.

Many have succeeded as technology entrepreneurs. They share a fundamental passion for, deep belief in, and total commitment to their activities. They are persistent, difficult to deflect from their goals. Most of all, we do believe everyone we know who has gone through this has greatly enjoyed their experiences.

Brian Culshaw and José Miguel López-Higuera
February 2008

Part I

Reflections, Motives, and Money

The main aim of this section is to provide a preliminary overview about the reasons and motivations for entrepreneurship and some comments concerning the key factors and their interrelation for the successful creation of new high-tech-based companies.

We can be thankful that we have evolved from the early stages, when the expectation of the man with the money was to find the entrepreneur's motivation basically centered on financial gain. For most, probably all of those involved with the adventures we encounter here, the prime motivation has been to innovate and see something useful emerge, while of course making—eventually—a decent income in the process. Emerging from the notoriously dreamy academic community are many modestly successful, and a few extremely successful, academic entrepreneurs who are responsible for a sizable fraction of new high-tech ventures. Awareness is increasing that academic institutions should find outlets for the diverse skills of their academic staff in much more than simply writing papers, preparing research proposals, and participating in academic-ego politics. The past 40 years have seen an immense broadening of the expectations on academic institutions. The major corporate industrial sector has likewise spread its scope and influence, with many major corporations spawning entrepreneurs from within their ranks and providing a source of the entrepreneur's often-friendly investment.

And what of money? Well, there are three basic approaches. The first is to use your own, though most of us don't have it. Earn to survive is the second and, while this may work for the small shopkeeper or self-employed carpenter, it is usually impractical for businesses of this nature, which require significant development funds to precede sales and returns. Usually, then, we all end up with someone else's, and the question is whose. The answer is complex. While this section of the book explores some of the options, the theme will recur. We shall see passionate advocacy for venture capital alongside entrepreneurs cautious of its influence and dominance. We shall see the role of local development agencies, private organizations, business angels, and bank loans. Some agreed to mortgage their own houses and livelihoods; others were more cautious.

The route an individual entrepreneurial venture takes is determined essentially by the motives of the people involved. There is a small chance of very large return with venture capital and there are those who successfully repeat this route using acutely developed business acumen, and making a fortune in the process. The high return success rate on venture capital investment is in the region of 10%. There are those who settle for much more modest, often frustrating, growth rates based on "private investments and then plough back the money" models. Some go through several stages and dip into each. Regional and national government-development aid is also an extremely important factor in virtually all communities in which high-tech entrepreneurship is encouraged—namely most of the developed world. The models for this vary immensely, from heavily subsidized accommodation to lavishly supported central facilities, to straightforward cash, to export development aid so often obvious as national and regional pavilions at major exhibits. And so it goes on.

This short section contains five contributions, many of which could equally well have appeared elsewhere, but which we selected on the basis of the message we perceived.

First, we have Guillermo Dehesa, who looks at reasons that determine success in high-tech organizations in this global world. From the analysis of the three main forms to achieve economic growth and the factors of success in developing high-tech companies, some key lessons are addressed and discussed. They include learning how to start, how to live in the new world of increasing returns, and how to finance a high-tech business.

The second chapter, by one of us (Culshaw), explores the academic as an entrepreneur, attempting to present the need for mutual respect and, possibly above all else, the necessity that the whole thing should be enjoyable. The third is Stuart Barnes' chapter on money. While it is presumptuous to extract themes from this or anyone else's contributions, parsimony and patience seem to feature strongly. Take note, too, of the final paragraph—one that is extremely important and all too easily put to the bottom of the list.

Mike Redman is an individual, also with a predominantly industrial career, who has been through a multitude of spinouts with varying success. Like most of our authors, he emphasizes the need for people to have trust and respect within a small team. Unlike most of our authors, he has exemplified this by relating personal experience with regard to what happens when this all-important trust breaks down.

Finally, we initially conceived Michael Lebby's chapter as addressing the role of his current activity—essentially trade association support. We see here a technical and philosophical adventure, and once more the emphasis on the need for fulfillment. Michael, like several of the technological entrepreneurs, has formally combined technical and business education in an MBA, though most have probably picked up their business acumen along the route.

Within these few pages we see the essential ingredients; motivation is critical. All entrepreneurs feel a complex "hunger" that is encapsulated in the urge to see something happen and push until it does. All acknowledge the absolutely essential role of the people around them and the necessity for a multifaceted team and partnership with others. And all need cash—from somewhere.

While certainly much more than these core elements is involved in the success of any entrepreneurial venture—and we shall see some of these evolve in the remainder of the book—we shall see these two basic themes: the motivations of the people involved and the need for financing.

1

Some Suggestions from an Economist to a New High-Tech Starter

Guillermo de la Dehesa
Chairman, Centre for Economic Policy Research (CEPR)
London, United Kingdom

In this chapter, the author looks at the factors that determine success in high-tech organizations around the world. From the analysis of the three main forms to achieve economic growth and the factors of success in developing high-tech companies, some key lessons are addressed and discussed. They include learning how to start, how to live in the new world of increasing returns, and how to finance a high-tech business.

1.1 On Learning How to Start a High-Tech Company

In any local economy or any business within the global economy, there are three main forms to achieve economic growth. One is through the accumulation and improvement of factor inputs such as labor and capital. Another is through developing trade and comparative advantage, and the third is through knowledge, innovation, and entrepreneurship. These three forms are not mutually exclusive and reinforce each other, but most advanced economies are increasingly specializing in promoting the last two, while most developing economies are mainly trying to develop and use the first two.

Knowledge, ideas, inventions, and entrepreneurs mostly tend to come out of universities and research centers. For instance, MIT faculty and students produce, on average, two new inventions every day, and there are close to 5,000 companies created worldwide by MIT alumni.[1]

What are the factors of success in developing high-tech companies? Experience shows that attitudes, talent management, patents, compensation, quality investors, speed, and location are the main factors.

Attitudes to innovation are essential, and that is what distinguishes the behavior of small companies compared to large companies. Radical innovations never originate with the market leader.[2] Even when the market leader developed the radical innovation, most times they will not pioneer it, often fearing that it would cannibalize sales of their existing products. Therefore, radical innovations tend to be created by small and successful start-ups (Microsoft is a case in point, but there are many more) because new high-tech entrepreneurs are never risk-averse; they have nothing to lose and everything to win. Nevertheless, there are countries and cultures that suffer both from the stigma of failure and from the stigma of success. It is paradoxical, but they always go together. These two stigmas have at their origin a common and fatally wrong belief: a zero-sum society in which the success of one is at the expense of another. If someone gets richer, then someone else must have gotten poorer.

Patents play a key role in creating a sustainable advantage for high-tech businesses. Radical innovations have much more leeway and possibilities than incremental innovations. The reason is that, for instance, when an entrepreneur with an innovation goes to look for a partnership with a company that is going to use that innovation and reduce its costs dramatically, the large company will react by thinking, "Do we really need these people?" Thus, if the patent position is weak, they will replicate it themselves; if it is strong, they will partner with you.

The quality of management teams is so important. Success tends toward a very strong management team with an average technology over a first-rate technology with a weak management team. Entrepreneurship is not about individual behavior but about team behavior. Significant empirical evidence shows that teams with complementary skills performed much better and made better decisions than individuals.

The form of compensation plays a key role in the success of the venture. High-tech start-ups using stock options as the main form of compensation tend to be more successful than others using a traditional way of remunerating with a cash and bonuses. The reason is that if you distribute ownership to the employees, they will behave as owners, they will no longer behave as employees, and they will face the same risks as the shareholders. The traditional "agency problem" between the principal and the agent or the shareholder and the executive disappears.

The quality of the investors is also a key factor for success. New high-tech ventures are costly and need more financial leverage than founders think—even more if they have any problem with the technology once they have started. The best investors are those who understand the innovation, believe in it, have deep pockets, and can leverage enough money. The time horizon of the investor is also very important: the longer, the better, given that once they leave through an IPO, the new shareholders may have a very short-term horizon incompatible with the diffusion of the technology.

The speed of the innovation to the market is another key factor. The quicker to the market, the higher the probability it will become a standard, being able to corner that market and get a good return from it. As the innovation cycle and product life becomes shorter, the quicker to market the better.

Finally, successful high-tech entrepreneurship needs a location or a "habitat" for the creation of new companies and new industries.[3] As mentioned earlier, although large corporations play an essential role in the economy by developing technological advances and new products, they are less likely to develop radical or disruptive technologies than can create major changes in industries or even start whole new industries. A clear example of this paradox is the creation of the biotechnology and information technology industries. For a start-up company, trying to develop an idea for a successful product is essential, while being at the center of the cauldron of ideas for new technologies and markets, represented by knowledge clusters such as Silicon Valley or other centers of agglomeration of knowledge, research, and development excellence.

The ideal habitat needs to have, in order to be really entrepreneurial and productive, the following features:

1. It should have the presence of universities and research institutions, where most ideas are created and knowledge is accumulated, because they have well-trained and experienced scientists and engineers.
2. It should have effective interaction between these universities and research centers and industry, given that their knowledge and ideas must pass between universities and industry, and vice versa. True intellectual engagement must exist between the two, so ideas and concepts co-evolve in a healthy two-way exchange. In order to achieve it, universities need to allow faculty to participate as consultants and advisors to companies, and companies need to be ready to sponsor research at universities.
3. It should have a high-quality labor force, where individuals and companies can find adequate manpower to start their high-tech ventures. In order to achieve it, the region needs to have an excellent education and training system with well-paid teachers, excellent facilities, and student support, where both public and private education can coexist. Such a labor force tends to have higher occupational and company mobility, which is also essential to collective learning in a community by moving their tacit knowledge (not secrets) from one company to another.
4. It should have a business climate that rewards risk-taking and does not punish failure. On the failure side, that means bankruptcy laws that provide limited liability to the invested capital and do not permit creditors to go beyond the availability of limited partnerships for venture capital firms. On the success side, it means security laws that bestow equity credit for knowledge, ideas, organization, and hard work so entrepreneurs have a better return than investors on their venture if it succeeds.
5. It should have an open business environment that promotes joint ventures and alliances between companies to develop further research, technology, and marketing. Some secrets tend to be more valuable when shared by several complementary companies.
6. It should have the presence of a venture-capital (VC) industry that understands high-tech ventures and knows that (in contrast with other start-ups) high-stake ventures that fail do not have residual value because their principal assets are intangibles: ideas, human knowledge, and technology.

7. It should have an open labor environment that gives a premium to diversity and does not discriminate against youth, gender, origin, and immigrants, and only emphasizes talent, merit, and individual and collective effort.
8. It should have a high quality of life in the habitat or cluster. Given that high-tech developers and workers are in huge demand globally, they have a high mobility and they prefer a temperate climate, good schools for their children, good health and recreation facilities, and comfortable housing.

1.2 On Learning How to Live in the New World of Increasing Returns

The business world in advanced countries has undergone a very important and diverse transformation in the last century: from one of dominant bulk-material manufacturing to another of dominant knowledge, innovation, and technology; from one of processing resources to another of processing information; from one using raw energy to another using ideas; from one of perfect competition to another of imperfect or monopolistic competition; and from one of decreasing returns to scale in the accumulation of physical capital to another of increasing returns to scale in the accumulation of human capital, knowledge, and ideas.

The idea of increasing returns can be found indirectly from Adam Smith's *An Inquiry into the Nature and Causes of the Wealth of Nations* (1776) in which he developed the idea of "the division of labor," where workers who engage in specialized, routine operations come to see better ways of accomplishing the same result, leading to inventions.[4] Later, Alfred Marshall, in his *Principles of Economics* (1890), developed the concept of "internal economies of scale," where internal costs are reduced but productivity is increased.[5] Conversely, "external economies of scale" are driven by external resources becoming available, which allow the firm to expand its operations.

Later, Allyn Young, in his paper titled "Increasing Returns and Economic Progress" (1928), used both concepts together: that of the division of labor, because it allows for transforming a group of complex processes into a succession of simpler processes and for being able to mass produce any product cheaply, and that of internal and external economies of scale, because by interacting with the division of labor, the progressive division and specialization, not of a single firm or industry but of all interrelated industries, is essential to increasing the returns.[6] Moreover, the division of labor depends on the extent of the market, and the extent of the market depends on the division of labor. Through this positive feedback among the two, increasing returns brings economic progress.

Finally, Nicholas Kaldor in his "Irrelevance of Equilibrium Economics" (1972) criticized the use of "constant returns to scale" in the models of general equilibrium, which produce a continuous equilibrium through time and of an exogenous rate of growth of the labor force and of capital.[7] He established that returns are not constant but increasing and that the forces making for continuous changes are endogenous (engendered from within the economic system). With increasing returns, change and growth become progressive and propagate and reinforce themselves in a cumulative way without limit. The development and econometric modeling of these ideas have been extended to competition by Avinash Dixit and Joseph Stiglitz (1977), to international trade by Paul Krugman (1979) and Wilfred Ethier (1982), to economic growth by Paul Romer (1987), and to labor markets by Martin Weitzman (1982).[8-12]

Increasing returns in the use of ideas, innovation, and technology produces a tendency for those who are ahead to get further ahead, and for those who lose advantage to lose further advantage. They are mechanisms of "positive feedback" that operate within markets, businesses, and industries reinforcing those who gain success and weakening those who suffer losses. Thus, increasing returns tend to generate winners and losers, and instability rather than equilibrium and convergence. They coexist alongside diminishing returns in the use of other factors of production in the present business world. While the latter holds in most of the traditional parts of the economy (the processing industries), the former reigns in the newer one (the knowledge-based industries).

Nevertheless, both kinds of business economies are mixed in reality. Manufacturing companies that process raw materials to achieve final manufactured products have operations such as design, branding, marketing, logistics, and distribution that belong to the knowledge-based world of increasing returns. This is the reason why many manufacturing operations are done in developing countries: to use labor more efficiently in labor-intensive products, while the knowledge-based operations are done in advanced economies; and to have access to technology and a high level of human capital. With the same logic, there are also knowledge-based businesses, such as computer software, that also are capable of manufacturing hardware, although most of the time they are kept in separate locations as well, based on the availability of human capital resources with lower relative unit costs.

Both kinds of businesses are worlds that differ in behavior, style, and culture, and they use different management techniques, strategies, governance, and codes of government regulation.[13]

With diminishing returns businesses (grains, livestock, heavy chemicals, metals, and ores) whose operations are largely repetitive day to day or week to week, competing means keeping products flowing, trying to improve quality, and getting costs down. In the last few decades, product differentiation, brand name, and economies of scale, scope, and diversification have allowed a new form of competition that is not based solely on cost and price, and that has evolved from "perfect competition" into "imperfect or monopolistic competition" where fewer very large companies dominate the market. But as they expand further, they tend to run into limitations in the number of customers who buy their brand or in the number of markets where they can be leaders or have easy access to raw materials. Consequently, no company is finally able to corner the market, because even if the products they make are different, they can be substitutable for one another, causing a standard price to emerge with margins getting thinner.

By contrast, in the increasing returns of knowledge-based, technology-driven businesses, competition takes the form of a "winner-takes-all or most" model in which an initial market lead can make it possible to dominate the entire market. The reasons for that different outcome are the following:

1. Because of the large size of their up-front costs, which increases the costs and barriers for new entrants to get in the market, computer hardware and software, pharmaceuticals, telecommunications, aircraft and missiles, bioengineering, etc., are difficult to design and deliver to the market. They are very heavy on know-how and innovation and light on resources, so they have R&D costs upfront that are very large relative to their unit production costs and only allow for very few companies with high levels of initial capital to enter and compete in the market.
2. Because of the network effects that are characteristic of knowledge-based businesses, many high-tech products need to be compatible with a network of users until they can become a standard, but the initial success of the product feeds on itself, and if it becomes a standard in the market, it will be very difficult to displace by any other new entrant.
3. Because of their "customer groove-in" tendency, high-tech products are typically difficult to use, requiring training, experience, and time. Once users invest in training, they need to update their skills for every subsequent version of the product. As more market is captured, it becomes easier to capture additional market shares because the opportunity cost and time spent moving to another competitive product.
4. Because of their "nonscarcity" quality of knowledge, the deployment of a physical asset in a specific task prevents its simultaneous deployment in other tasks. Knowledge assets are nonscarce because they can be deployed simultaneously in multiple or infinite tasks, reducing costs and enlarging customers. In sum, knowledge-based businesses have high costs of entry and, at the same time, very large economies of scale and scope, which tend to make the first product the standard in the market and extremely difficult to displace (but not impossible).

There are also many other differences in favor of knowledge-based businesses versus traditional processing-manufacturing businesses. The first difference is in their management structure and corporate governance. Processing bulk resources through repetitive operations favors an environment free of big surprises, which allows for careful control and planning on the one side and permanent cost optimization and quality improvement on the other. Both trends tend to favor a hierarchy between workers and bosses, between producers and controllers, and planners and production optimizers.

By contrast, in the knowledge-based businesses, which are competing in a winner-takes-most market, managing is about finding the next technological winner, the next big thing, or the next "cash cow." Management becomes mission oriented, not production oriented. Hierarchies flatten because, in order to deliver the next big thing, they need to get organized as "commando" units that report directly to their CEO, and the unit teams are treated not as employees but equals who all are key elements of the company's success. Thus, this kind of business demands flat hierarchies and mission orientation, not five-year plans.

The second difference is the in competition. In the decreasing-returns business, the aim of competition is to be able to keep high-quality products flowing at a low cost, and there is no need to watch the market every day. By contrast, in the increasing-returns business, the style of competition is similar to "casino gambling," where one part of the game is to choose which games to play and another part is to play them skillfully enough to win. Psychology also becomes very important because you need to bluff your potential competitor to make him think that you have already locked in the market or that your technology is better adapted to the market demands.

For that, you need to have not only a high technical expertise, but also deep pockets and a lot of courage because you do not know or cannot totally guess which technology or innovation is going eventually to become the winning one. Because the game is not fully defined, it is very difficult to optimize. It may be better to adapt and watch for the next wave that it is coming in order to position the company to take advantage of it. Thus, adaptation often means fully reinventing your company for the next wave, which would be unthinkable in a bulk-processing industry.

With this gambling-like competition it pays to hit the market first, if possible, starting with low, attractive prices; it also pays to have superb technology to maintain your leading position. But these two do not fully guarantee success. You can be displaced by a competitor with deeper pockets and similar technology that gives away its technology in order to capture your clients. Once that product becomes the standard, the company asks for payment for every incremental improvement in the technology in order to get a decent return.

To achieve a locked-in product, it is necessary to have a high degree of convenience, be very simple to use, and very fast to deliver (time is money). It needs to have a fair price and should not obstruct technological advancement; thus, every competitor needs to have a level playing field. The temporary monopoly of a locked-in product must be seen as a reward for its contribution to innovation and to the risk taken in its research and development.

The third difference is that technological products do not stand alone, but they depend on the existence of other products and other technologies that support and enhance them. Thus, they exist in interdependent "mini-ecologies" in which you need to have more or less loose alliances with other companies organized around a determined mini-ecology. It is there that your new product is going to operate; otherwise, it will be extremely costly and difficult to create another new, successful product without the help of the other members of the mini-ecology.

1.3 On Learning How to Finance a High-Tech Business

Another important difference with knowledge-based assets and products is that they create significant "information asymmetries." Sanford Grossman and Joseph Stiglitz were the first two economists to argue that efficient financial markets are not possible, from the point of view of information, because efficiency would prevent equilibrium.[14] According to both economists, the "efficient market hypothesis" determines that prices reflect all available information and that information has a cost, and this would actually lead to the collapse of competitive markets.[15-17]

Under this hypothesis, every informed investor in a competitive market feels he can stop paying for information and be as efficient as another investor who is paying for it. But all informed investors feel the same need. The complete lack of informed investors is not an equilibrium situation since each investor, by taking the price as given, feels he can gain greater profits by being better informed. The efficient markets hypothesis states correctly that costless information is a sufficient condition for prices to reflect all necessary information, but it does not state, too, that it is also a necessary condition. Nevertheless, this is a "reduction ad absurdum," since price systems and competitive markets are important only when information has a cost.[18]

Under the efficient market hypothesis, then, equilibrium is only attained when information has a very low cost or when informed investors obtain very precise information. But since this information has a cost, prices cannot reflect all the information that is available because, if they did so, those who pay for the information would receive no reward. There is, consequently, a fundamental conflict between the efficiency with which markets distribute information and the incentives that exist to acquire it.

These problems of "asymmetric information" derive from situations in which one of the parties to a financial transaction has less information than the other.[19-20] This tends to give rise to an inefficient allocation of financial resources, which leads to "agency problems," "adverse selection," "moral hazard," and "herd behavior." These kinds of problems are applicable to lending banks, investors, and fund managers, and are defined as follows.

Agency problems, or principal and agent problems, arise when, for example, shareholders entrust one or more executives to run their company for them, or when small shareholders invest in a fund managed by experts.[21] In such situations, the information available to the shareholder or investor is far inferior to that available to the executive or fund manager. As a result, it is very difficult for the former to know whether the latter is correctly performing the task entrusted to them. If the preferences of one party differ from those of the other, the result will be less than optimal.

Adverse selection often occurs when lenders have incomplete information on the quality of borrowers.[22] This can provide the worst, or riskiest, borrowers with more incentive to request loans. When lenders cannot obtain sufficient information on the credit quality of borrowers, they try to apply price or interest-rate conditions that reflect the average quality of the relevant borrower group as a whole, which can be adverse for higher-quality borrowers and beneficial to higher-risk borrowers. As a result, the higher-quality borrowers may curtail their operations in such markets, and the lenders may end up lending primarily to the riskiest borrowers, causing the resulting allocation of funds to be inefficient.

Moral hazard can occur if a borrower's behavior changes in a way that is undesirable for the lender after the transaction has already taken place.[23] For example, a borrower may be inclined to use a loan for a relatively risky project and, if everything works out well, the borrower will be successful. If it does not, it is the lender who stands to lose. As a result, the lender is inclined to make sure the risk on the project is minimal. To bypass this, the borrower may try to alter the project, making it riskier once the loan has been approved. As a result, the project may turn out to be much riskier than the lender had anticipated, making the lender reluctant to grant loans in future and, consequently, his loan volume will end up being less than optimal.

Finally, herd behavior is characterized by an increasing inclination by lenders and investors—due to a lack of information—to follow those who they believe have more information on a certain borrower or investment opportunity.[24-25] Such behavior can cause markets to experience brusque movements and high volatility if a lender, considered by others to have better information on a borrower, ceases, for whatever reason, to lend money to the borrower, prompting others to do the same and driving the borrowing company to default, even though the borrower's solvency situation was not really a concern.

Such behavior also arises when investors do not have enough information on the quality of their fund managers. The fund managers with the least quality will have an incentive to emulate the investment decisions of other managers they consider to be of higher quality to "hide within the herd," so it is not as easy to assess their skills. Even good fund managers may have reasons to follow the market because if they go against it and they lose, they could risk losing their jobs, while if they follow the market and everybody loses, they are safer.

As I said at the beginning, high-tech projects can create a lot of information asymmetries.[26-27] The first cause is that high-tech projects, by definition, have a large uncertainty and risk associated with them. For instance, the prospects associated with the development of a new drug are much more uncertain that a tangible real-estate project. Investors react to uncertainty by demanding a higher return or premium, which translates into a higher cost of capital to the company. An excessive cost of capital may hinder the new investment or force a viable high-tech project to be abandoned.

The second cause is that if it is already difficult for the lender or investor to know and understand the tangible projects that are presented to him. In the case of high-tech, knowledge-based intangible projects, it is almost impossible unless they have the advice of high-tech scientists, and even then they cannot be sure about the risk they are taking since only the borrower knows better. Thus, again under asymmetric information, the investor tends to charge a higher interest rate or achieve a higher return than with simpler products or projects.

A third issue is that the accounting authorities still use archaic accounting rules that treat most investments in knowledge as regular expenses instead of as investments, so they require the immediate expensing of all costs. This depresses reported earnings, making them less profitable and more undervalued. The outcome of such deficient accounting is deterioration in the information of reported earnings, which is the prime product of the accounting system. Less informative earnings obviously decrease the investor or lender knowledge and increases the effects of information asymmetry.

Therefore, although investors generally recognize the primacy of knowledge assets as value creators, they do not count on capital markets to value them properly, so their return has to come from a higher cost of capital to the borrower.

For these reasons, high-tech start-ups must be financed through specialized venture capital, either by VC funds or by venture capitalists, and not by traditional banks. Banks are incapable of adequately financing innovative firms and, in particular, high-tech start-ups.[28-29] Venture capitalists prefer to invest in young and innovative companies or start-ups because, although they have a higher risk, they expect to get higher returns in the future. Since they are specialized in this kind of financing and also act as monitors in related firms, each new investment not only lowers their cost of monitoring but also generates externalities that can be used in assisting and mentoring other future start-ups. Their specific technological expertise generates higher-margin returns compared to unspecific financiers and traditional banks, which prefer to lend based on tangible collateral rather than risky equity without collateral against only the guarantee of the project.

One of the reasons VC has developed much less in continental Europe than in the U.S. or the UK is that most continental European financial systems are universal-banks based instead of capital-markets based. Capital markets allow for diverse views from different investors: some are ready to take more risk than others; some prefer to invest a small part of their portfolio in high-risk assets or projects; while banks tend to be very cautious of lending without collateral and try to find an internal consensus before lending. Large, efficient, and liquid capital markets offer fertile ground for institutional investors (investment banks, insurance companies, investment funds, hedge funds, and pension funds) to invest part of their funds in VC funds and companies.[30]

There are two main reasons for the reliance on capital markets. First, is that VC investors are powerfully attracted by the possibility of exiting from their investment through an initial public offering (IPO), which allows both investors and managers of VC backed companies to maximize their return on investment. Second, in most capital-markets-based systems, there are powerful stock exchanges devoted to high-tech and innovative companies separate from the main exchanges, which attract these kinds of companies to be quoted as well as specialized investors in these kinds of companies (NASDAQ started in 1971, and the European New Markets started in the late 1990s). Nevertheless, exiting through sales to other VC investors or to the start-up's management (after a few years) is another alternative and also a very frequent form of quitting.

VC may be more expensive than other forms of finance, but that reflects having an involved and specialized investor who takes care of the project, gives strategic advice, and monitors it. Friend and family finance should only be viewed as complementary for venture capital rather than as a substitute; but in any case, the most successful start-ups are those initially owned by the developer of the idea, not by friends and family.

Many years of experience with VC high-tech business financing shows that venture capitalists take the following factors into consideration: First, the better the intellectual property of the start-up (measured by the quality of its patent), the easier it is to get financed by a VC. Second, the higher the human level of capital of the high-tech starters, the higher the likelihood of getting VC financing. Third, the higher the ownership share of the starters, executives, and other friends and family, the lower the likelihood of getting VC financing because it shows that they have not been able to get financing from outsiders. Fourth, the larger the amount of debt accumulated, the lower the likelihood of obtaining VC financing. Fifth, the higher the likelihood of a compensation system through stock-options (rather than by cash), the higher the likelihood of getting VC financing.

References

1. Preston, John T., "Success factors in technology-based entrepreneurship," MIT Entrepreneurship Center (August 2001).
2. Utterback, James, *Mastering the Dynamics of Innovation*, Harvard Business School Press, Harvard University, Cambridge, MA (1994).
3. Miller, William F., "The habitat for entrepreneurship," Stanford University Graduate School of Business, (July 2000).
4. Smith, Adam, *An Inquiry into the Nature and Causes of the Wealth of Nations*, Modern Library Eds. Random House, New York (1776 and 1937).
5. Marshall, Alfred, *Principles of Economics*, MacMillan and Co., London (1890 and 1961).
6. Young, Allyn, "Increasing returns and economic progress," *The Economic Journal*, Vol. 38, No. 152 (December 1928).
7. Kaldor, Nicholas, "The irrelevance of equilibrium economics," *The Economic Journal*, Vol. 82 (December 1972).
8. Dixit, Avinash K. and Stiglitz, Joseph E., "Monopolistic competition and optimum product diversity," *American Economic Review*, Vol. 67, No. 3 (June 1977).
9. Krugman, Paul R., "Increasing returns, monopolistic competition and international trade," *Journal of International Economics*, Vol. 9, No. 4 (November 1979).
10. Ethier, Wilfred J., "National and International returns to scale in the modern theory of international trade," *American Economic Review*, Vol. 72, No. 3 (June 1982).

11. Romer, Paul, "Growth based on increasing returns due to specialization," *American Economic Review*, Vol. 77, No. 2 (May 1987).
12. Weitzman, Martin L., "Increasing returns and the foundation of unemployment theory," *The Economic Journal*, Vol. 92 (December 1982).
13. Arthur, W. Brian, "Increasing returns and the new world of business," *Harvard Business Review*, Vol. 74 (July-August 1996).
14. Grossmann, Sanford and Stiglitz, Joseph E., "On the impossibility of informationally efficient markets," *American Economic Review*, Vol. 70, (June 1980).
15. Fama, Eugene F., "The behavior of stock markets prices," *Journal of Business*, Vol. 38, No. 1 (January 1965).
16. Fama, Eugene F., "Efficient capital markets," *Journal of Finance*, Vol. 46 (1991).
17. Roberts, H., "Statistical versus clinical predictions of the stock market," University of Chicago, (May 1967).
18. Hayek, Friedrich A., "The use of knowledge in society," *American Economic Review*, Vol.35 (1945).
19. Spence, Andrew M., *Market Signaling: Informational Transfer in Hiring and Related Screening Processes*, Harvard University Press, Cambridge, MA (1974).
20. Stiglitz, Joseph E. and Weiss, Andrew, "Credit rationing in markets with imperfect information," *American Economic Review*, Vol. 71 (1981).
21. Grossman, Sanford and Hart, Oliver, "An analysis of the principal-agent problem," *Econometrica*, Vol. 51, No. 1, (January 1983).
22. Akerloff, George, "The market for lemons: qualitative uncertainty and the market mechanism," *Quarterly Journal of Economics*, Vol. 84 (1970).
23. Arrow, Kenneth J., *Essays in the Theory of Risk Bearing*, Markham Publishing Company, Chicago, IL (1971).
24. Gwynne, Sam C., *Selling Money*, Weidenfeld & Nicholson, New York (1986).
25. Banerjee, Abhijit V., "A simple model of herd behavior," *Quarterly Journal of Economics*, Vol. 107, No. 3 (August 1992).
26. Lev, Baruch, *Intangibles: Management, Measuring, and Reporting*, Brookings Institution Press, Washington DC (2001).
27. Lev, Baruch, "Knowledge and shareholder value," *Stern School of Business, New York University* (January 2000).
28. Gompers, Paul A. and Lerner, Josh, *The Money Invention: How Venture Capital Creates New Wealth*, Harvard Business School Press, Harvard University, Cambridge, MA (2001).
29. Audretsch, David B. and Lehmann, Erik E. "Financing high-tech growth: the role of debt and equity," Max Planck Institute of Economics, Group for Entrepreneurship, Growth and Public Policy. (2004).

30. De la Dehesa, Guillermo, "Venture capital in the United States and in Europe," *Occasional Paper No. 65*, The Group of Thirty, Washington DC (2002).

2

The Academic Entrepreneur: An Oxymoron?

Brian Culshaw
Professor, University of Strathclyde
Glasgow, United Kingdom

This chapter explores the academic as an entrepreneur, attempting to present the need for mutual respect and, possibly above all else, the necessity that the whole thing should be enjoyable.

2.1 Introduction

The concept that academics should "take their ideas to market" has been with us for a very long time. Historically, though, mutterings in the common room have disparaged the (usually) senior academic trotting the world to his planning consultancies in Tokyo, London, Cape Town, and New York. How can such philandering possibly benefit the academic credibility of our august institution? Indeed, why are such miscreants never discreetly asked to resign from their positions?

Similar gossip survives to this day. However, in the UK at least the encouragement to do the deed as well as publish the paper has never been stronger. Universities gain local credibility through reporting that their staffs have successfully taken this or that high-technology concept into a marketable product. The university "contracts office" has evolved from a quirky curiosity into a (usually) professional support body, linking academia into sources of finance and industrial exploitation opportunities. Institutional philosophy may have changed, but the academics themselves retain many of their stereotype attributes, dreamy and unmanageable ("herding cats" is the phrase often used); they are anathema to conventional business. Spin-outs, indeed the whole SME (small-to-medium enterprise) culture, shouldn't, though, be confused with the conventional business infrastructure embraced by the corporate multinational.

2.2 Why Should Academics Get Involved?

The vast majority of academics enjoy tenure, the prospect of a reasonably secure pension, the freedom within very broad limits to do as they please (especially in the research context), and an idealistic genteel lifestyle, intruded upon only by the frenetic excesses of research-assessment exercises and quality-assurance agencies. The motivation then is certainly not basic survival. Neither is it the desire to accumulate a fortune. Academics are notoriously disinterested in large sums of cash, provided, of course, that the wine cellar is well stocked. Engineering academics are a somewhat different breed: scientists who when young tinkered with construction sets, and so like to see ideas turned into reality. After all, in academia the definition of reality is the published paper. The grown-up construction kid often wishes to see more.

So, personally, I started all this to see things happen. We would find things out in the lab, publish the paper, and be left with this lingering thought that really this might be useful somewhere. Taking this precious thought to the corporate multinational is almost certainly doomed to failure. The said corporate servants convincingly argue that they have already had a magical product on the market for the last 15 years doing exactly what you claim your nebulous thought might be able to do. Though, we discovered that the users of this magical product may well beg to differ. In time, the inevitable conclusion is the only way to push your baby is to push the pram yourself. Consequently, the thought of a corporate vehicle through which to make things happen eventually consolidates into an actual entity.

There are other motivations as well, though these tend to appear in hindsight. The spin-off can, and often does, offer a different type of opportunity for the right type of emerging graduate student, and that in itself can be exciting and rewarding. Furthermore, the direct experience of actually making things happen is splendidly effective at bringing the dreamy and unmanageable academic rapidly down to earth. The apparently simple step of putting the lab demonstration into a box and shipping it out the door raises a whole raft of absolutely essential and very demanding technical issues that are all too rapidly trivialized by those who have escaped but observed the process from afar.

So in the late 1980s, a few of us in the university, in collaboration with a local company and our regional development agency, set up Gallex with essentially these ends in view. We got investment, hired a managing director, took management advice (after all, we are academics and don't know how to manage). Two years later, Gallex was wound up. In the UK there is a government enquiry into corporate bankruptcies, the result of which was that our professional managing director was banned from being a company director for several years. For us, the trouble with Gallex was that it never felt right. But the experience of bankruptcy is frowned upon in European cultures far more than the U.S. Lots of musing later, along came the conclusion that we should trust our instincts. Perhaps intuition after all is really quite a good thing.

We regrouped early in 1994. Two of the academics from the original Gallex team, another academic from a neighboring university, and the former chairman of a company with whom we at Strathclyde had done some successful technology transfer got together, and out came OptoSci Limited. OptoSci was founded with the same objective, namely to take things from the lab into reality.

We had by this time learned that there are two basic models for a spin-out company. The first is to take an initial modest investment, including that from the founders, and earn your keep. Remember, too, that in Europe there is no equivalent of the rather splendid U.S.-based small-business innovative research (SBIR) scheme. This simply contravenes the EU equivalent of antitrust legislation, so earning and simultaneously developing a new product within a SME presents a challenge. Of course, this challenge can be met by persuading the venture capitalists or possibly the business angels (many of whom respond to the name "Lucifer") into the fold. This, though, loses control and intuitively—based on the Gallex experience—control of one's own destiny is, at least in the early stages, a critical prerequisite. Consequently, we went for the modest-investment-and-earn model for OptoSci. A direct result of this is that growth is probably more modest, but total collapse is less likely.

The key to earning your keep is a readily developed, easily marketed product of reasonable cost that can earn sufficient profit to be able to plough useful amounts, with matching government funding, into development programs. In our case, we opted initially toward laboratory instrumentation for the undergraduate teaching market. OptoSci's resulting range of photonic educator kits is unique in addressing highly technical concepts within a reasonable cost envelope, with the benefit of both excellent supporting documentation and significant student exposure prior to market release. Additionally, the founders are predominantly academics who know how academics tick and how the funding cycles in academic institutions work, so we can time offers to potential customers with the tantalizing prospect of money to spend. This, we later learned, is called "marketing." Of course, the market is limited and also needs to be very multinational to become viable. We then had to learn about foreign agents, letters of credit, and the whims of exchange rates.

The model though has served us well and, in effect, the education market has helped us provide matching funding for government initiatives, targeting more speculative products. In particular, we have begun to address gas sensing using fiber-optic systems and have now installed systems, some of which have been operating for several years, in very challenging environments, particularly for the detection and accurate measurement of methane gas concentrations.

We have, though, continued along the modest-investment-and-earn model, despite the temptations at the beginning of the millennium when photonic technology was the venture capitalists ideal. Thankfully, we resisted the temptation and have survived. The model, though, has its frustrations of which the demands on company staff are undoubtedly the greatest. The need to survive is paramount—true of any company, large or small, so longer-term prospects often incur frustrating delays. We do, though, retain that ever-precious control, and with the company being consistently profitable for several years we strengthen our position should investment be required in the future.

What of the future? Educational equipment is a stable product and has been augmented in recent years by a range of instrumentation products and OEM units. The large systems for safety monitoring, initially focusing on methane, appear particularly promising, and we are working on expanding the market address. However, we really don't know how things will turn out, so flexibility and agility—coupled to sound, even parsimonious financial control and carefully considered forecasting—are always on the agenda.

2.3 And What Have We Learned?

Taking ideas from the academic paper to a real product has far more to it than we first thought. In particular, we found the need to devote far more effort to persuading people to buy our product than we do to developing it. The illusion that "it is technically excellent so it sells itself" soon vaporized. For many high-technology companies, certification of product (administration and form filling—another anathema to academics) is absolutely essential, and the whole issue of putting the idea into the right kind of box that survives in the right kind of environment, under demanding conditions, is extremely important. The glue that holds together that critical assembly is probably the most closely guarded secret of the entire process.

People too are critical. The SME is a small community. Everyone interacts daily. Staff motivation and staff selection are therefore vital elements. Thus far, we have largely employed people whom we have known, either personally or through reliable reputation. There is a need for the more nebulous benefits, too. Share options for key staff (even though we may never sell the company) give an essential sense of belonging. Most of all, everyone concerned must trust these nebulous feelings and instincts and have faith in intuitive reactions. Later, we may need to logically justify our choices and actions, but if it doesn't feel right, don't do it until it does. The SME can very productively employ eccentrics and unusual characters, making the best of technical excellence and working with nonconformism. Paradoxically, perhaps the SME can have overtones of the academic environment that often stimulated it, so perhaps it isn't so alien after all.

Going back to the university, there are still tensions and often the academic employer and the spin-out company can pull in opposite directions. At one extreme, in my head-of-department era, one senior member of staff appeared to announce his successful acquisition of venture-capital funding and that, by the way, he would be leaving in two weeks on an extended leave of absence. Incidentally, it was all fine since the university centrally had approved this, but for commercial confidentiality reasons the department couldn't be informed of the negotiations. But it was then up to the department to sort out all the teaching, administration, and research duties associated with this particular staff member. No further comment needed!

Spin-offs often intrude in other ways. University lab equipment and space is trespassed upon by company staff, who are often given supreme priority. Yes, unfortunately, there are cases where the spin-off dominates and departmental resentment festers. Fortunately, though, such cases are relatively rare since academics have learned the dangers.

There are benefits, too. OptoSci has supported graduate students, loaned equipment, even built the occasional special piece of kit at cost or less. The academic staff concerned has gained immensely in experience and bring an authority to undergraduate teaching and research that cannot be achieved through other routes. At the institutional level, there is this somewhat altruistic contribution to the local community, which itself has immense value. When a university spin-off is handled carefully, these and other benefits far outweigh the frustrations.

None of us involved has made a fortune, and it is unlikely that we will. Would we go through the same exercise again? I am sure the answer would be most definitely in the affirmative. Indeed, in my own case recently appeared Solus Sensors, this time based in Wales and so hardly a campus company. But, again, this is targeted at taking a particular sensing system, well proven in the laboratory, and turning it into a useful product. We have already done some preliminary assessments in collaboration with a potential customer. I have no doubt the story will continue.

2.4 So Is the Academic Entrepreneur an Oxymoron?

We know for sure that our business has survived over a decade, in itself a tangible measure of success. We have had the satisfaction of really turning our eccentric research into a quality product. We have learned of the intangibles that never appear in the books, courses, and consultants' reports. We have valued, and will continue to do so, the immense contribution from energetic, capable, and dedicated staff. Perhaps most of all, we've learned that instinct and persistence make a powerful combination. All the academics involved would, I'm sure, repeat the exercise with the benefit of a more mature judgment on the process. We also know that our institutions perceive sufficient value in such initiatives to be willing partners.

3

Money

Stuart Barnes
Entrepreneur and Consultant
London, United Kingdom

In this chapter, the role of money in the entrepreneurship process is analyzed, including conclusions that parsimony and patience feature prominently in entrepreneurship.

3.1 Introduction

Money. This may seem a strange topic for a chief technical officer (CTO) to address, particularly one so unversed in financial jargon! However, as CTO you will probably be the one most responsible for blowing the company's cash, particularly in the early stages when you are proving all those technological ideas (sold eloquently to venture capitalists) and subsequently engineering them into a qualified product suitable for your chosen market.

First, let me put in a word of support for the much-maligned venture capitalist (VC). We should not forget that these people are looking after someone else's money—more often than not, large chunks of money from "super investors" such as insurance companies. But down at the bottom of this food chain are individuals: you, me, your grandparents. VCs have fiscal responsibilities, and most discharge these very well. Returns from this source out-perform the stock market by at least one percentage point over the long run. I get pretty enraged when I hear them called "vulture capitalists." Typically, they aren't, and the best work hardest and most conscientiously when they are trying to save a struggling company. The individuals making these observations seem to be the people who are too afraid to leave the safe surroundings of bigger companies or those that leaped in when the start-up industry was apparently printing money.

Perhaps the only generic criticism I have of VCs is their inflated claims about added value. A few years ago, they claimed to have all the skills under the sun: recruitment, patents, and so on. Don't believe it. If you have plenty of miles on the clock—as both I and surprisingly many start-up entrepreneurs have—then you probably have much more real experience in these areas. Happily, this fashion appears to be dying a quick death, and VCs are getting back to doing what they do best—providing equity to start-ups.

3.2 Getting Hold of the Money in the First Place

First, there seems to be two predominant models. European VCs tend to look at technology as the pivotal component of a start-up. Many European VCs actively work with university business groups to get in early! This means that technology plays are more typical in Europe both in telecommunications and life sciences, with relatively few (compared with North America) in Europe bucking this trend. A further consequence of this is that chip sizes are lower. Exit valuations scale accordingly. So European VCs appear to be happier building a business out from a piece of technology. This may be a self-reinforcing cycle or simply due to the fact that Europe is a generation behind the U.S.

Interestingly and paradoxically, U.S. VCs appear to come at it from a different perspective. Yes, there are lots of good ideas coming out of Stanford, MIT, Purdue… but the U.S. model, which is notably more seasoned, tends to look at teams, arguing (again fairly reasonably) that a good team will find (or has found) the right technology and has the skills to execute. From this position, they seem more comfortable taking bigger bets (for instance, big-box plays such as Juniper or Cisco). These would find it harder to get going in Europe as the land lies presently. Furthermore, U.S. VCs are much more tolerant of failure because they understand the odds; one in 10 start-ups will be a stellar success, two will do OK, and seven will bomb. Whether this difference between U.S. and Europe is a timing question or a cultural issue will only be resolved over the next couple of decades as the European market matures.

3.3 And…

I make no claim to understanding the intricacies of accountancy, subscribing to Micawber's (from Charles Dickens' novel *David Copperfield*) approach to happiness from a startup, rather than domestic perspective. However, the financial model is fundamental to the whole thing getting off the ground. You will be pitching to financially savvy VCs. What they don't know, they will fill in during subsequent due diligence. If you don't understand the terminology, find someone who does.

What are the fundamentals? First, carefully scope what it costs to develop the product to the point of sales. It's amazing how many technology start-ups think they need millions for their pet science project and fractions of that to engineer it into a product. Of course, I am an engineer and would say that, but I do have the scar tissue, too. Be prepared to be in it for the long run. Milton Chang—one of the best, having built two successful businesses—believes that it takes 10+ years to build a successful company. I believe it should take less, but not a lot less. It takes time, and there is a cost associated with this. Don't get rolled over on this because it conflicts with the model used by the VC, who will want to be in and out in seven years max to satisfy the investors—hence the dilemma. If you do get rolled over, you could end getting thrown out or washed out by VCs as you continually miss financial milestones later in the company's life.

The second fundamental is the product itself. What are you selling? The technology itself, or products emanating from the technology? Whatever, you need to understand the marketplace (market size, £1M per annum; cost of development, £10M; result, misery). You will need sales to get the company away, unless you were one of the lucky ones who sold the company on a bunch of PowerPoint slides in the late '90s. If you have this deluded impression, think again. Companies won't be that stupid this time around (or at least don't plan for it). Know your market. If necessary, pay for a market analysis, even though the VCs will discount it. It is better than nothing, which is what you will get. As with the engineering element, don't forget there is a cost to sales. Good salespeople cost and will want a slice of the action.

If you have scoped these things, you are well on the way to crafting the business model and nearly ready to pitch for the "wonga." Once you have this, talk to as many people as you can. "Free-ish" advice is surprisingly easy to come by. There are at long last some serial entrepreneurs in Europe. Run your ideas past them. Be brave. Also, VCs will more often than not throw away presentations from "unknowns." Try to get the advisor to sponsor the idea. Perhaps the best source of advice is from the VCs themselves. They may not like your idea when you are pitching, but they often give feedback. Listen and take notes. This will sharpen you up for future presentations. It's very easy to become discouraged when you know your pitch hasn't hit the spot—and it doesn't, often. Don't be afraid of rejection. There are lots of hidden reasons other than that they think the idea stinks (which, of course, they might). The fund may be fully allocated, there may be a similar company in their portfolio (which becomes evident during the presentation), the chemistry doesn't work, or any number of other reasons. Use the presentation time and ensuing feedback to improve the next presentation.

Finally, don't even start if you think this is easy street. That, it ain't, so why put yourself through the wringer?

3.4 Hanging on

Getting the money is hard enough. Hanging on to it is a different thing. It's amazing what effects suddenly getting a large amount of money can do to once-rational people. Seasoned professionals from big organizations, accustomed to brutal annual budgeting regimes, suddenly lose their marbles. Academics previously used to "paper and string" research suddenly have more money in an accessible account than they saw in decades previous. I knew of some U.S. academics who, having once been used to economy class and budget hotels, suddenly deemed that they could only travel first class and stay in the swankiest hotels at several hundred dollars a night. The vanity of it all.

Right from the get-go, you should be in firm control of the finances. There are a range of professional bloodsuckers out there that can sniff out that heady mixture of cash and inflated ego. And if this is not sufficient to make you careful, bear in mind this is not your cash to spend at will. There is a vulnerable pensioner somewhere at the end of the chain who deserves your financial prudence. It may even be one of your loved ones.

There are lots of "professionals" out there willing to take their share of your hard-won cash. It starts at day one with the legal profession. Lawyers are a necessary evil and add value, but make sure that you have a budget and shop around. Don't forget that out-of-town rates may be less expensive (25% of London rates, for example).

Patents are also necessary and becoming even more expensive. Find a good attorney and identify a low-cost process that suits you. National filings give you a rough-and-ready search and a filing date, followed by a patent cooperation treaty (PCT), and then, as late as possible, the various specific country filings. This allows maximum time to determine whether the patent is viable or not. Don't be too proud to admit that it isn't such a good idea. Hanging on can cost you greater than $10,000 per patent.

Where to locate is the next issue. If you are lucky enough to be in a designated regional development area, take advantage of it. If not, take time to shop around and find suitable premises. Negotiate hard. If you don't ask, you don't get; and this is a recurring cost and influences your shut-down costs. Get it wrong and you pay yesterday, today, and tomorrow.

Funnily enough, I've also found VCs strangely equivocal about this. It may be that from their palaces in Mayfair (an upscale part of London) they cannot see the harsh reality of this element of building a business.

Fitting out the facility is another honey trap. Shop around; buy benches and furniture at auction if possible. In the technical sector, it's hard to do without substantial amounts of test equipment, but be careful. It's hard to not succumb to fearmongering tactics such as, "If you don't buy this, there won't be another available for months." When you are up against time pressures, it's easy to leap in. Stay cool at the moment the supply/demand balance is in the purchaser's favor. Negotiate hard on price, and bear in mind that lease-purchase deals may be more favorable from a cash-burn perspective. Don't be afraid of the second-hand market and auctions, including the online versions. You can get an extra sense of pleasure if you have purchased lots from competitors!

Similar pressures can also be applied by raw material and component suppliers. At ilotron, my previous start-up, we were more or less pressured into buying a million dollars' worth of a particular component because it was that or nothing—even if we needed no more than $50K worth! It was a time when component suppliers were practically printing money, so why drive a potential long-term customer to the wire? Who knows? We have experienced similar situations with Azea, interestingly again from a U.S. West Coast player, so cultural issues are probably a factor here. My instinct now is to look east, where we have encountered nothing but reputable suppliers.

One further tip: negotiate hard on that first price. This is the benchmark for the future. Better still, ask for samples and avoid setting that initial price.

I've saved the worst for last—headhunters, who in my opinion are below estate agents in the food chain. There are a few good ones out there, but down at the bottom of the barrel it's pretty grubby. In a previous life, one particular headhunter got his hooks into a young engineer of mine, tapping him mercilessly for information, which he traded vigorously inside and outside the company. When the lad lost his job, did this guy ride in to find him a job? You guessed it—he'd got his poisonous barbs into someone else and was not around to help his erstwhile "friend."

Finally, stay on agenda. Cash really is king. Don't let it slip away. Once it is gone, none of the above will help you get it back. Don't forget, it is not your money; you are investing it for someone else, and if you are successful you will get a healthy dividend. Don't fritter it away.

3.5 And Forget...

Money from the Department of Trade and Industry (DTI) and European Economic Community (EEC). Though I'm sure these bodies have noble intentions, they are so riddled by bureaucracy that by the time they make a decision you could be at the end of your funding cycle, or worse still, finished!

These organizations (and I'm sure I'll get howls of protests here) are much better geared to helping out bigger corporations, where bureaucracy meets bureaucracy. From my perspective, I'd be much happier to see the money go straight to universities and forget "speed dating" and the rest of the rubbish. Then, eliminate the full economic tariffs on small companies who find academic partners. Then, let market forces do their work.

3.6 A Nonfinancial Footnote

I make no apologies for concentrating on money matters, as this is the nub of making your dreams reality. However, as this is supposed to be a personal impression, I'd like to offer these extra words of wisdom, which, bearing in mind that I have not succeeded to date, you may choose to disregard!

First, make sure a start-up is the thing for you. You must have a passion for either the technology or the environment. Don't go for it if you think you are going to get rich quick: one in ten start-ups is a blinding success, two in ten do OK, seven fail. Not all these failures have been bad ideas or badly executed. For many, it's just been bad timing or bad luck. Many of the blinding successes are also simply good luck!

Second, a comment on the speed factor: as mentioned earlier, Milton Chang, one of the most respected gurus in our industry believes that it takes 10+ years to build a business, and he has successfully built two. This may be a little too long for most of us—including the VCs—but it is a salutary message for all those looking to get rich quick.

Don't plunge in with the first VC who waves a term sheet at you. If one thinks it's a good idea, others generally will too, and this will enable you to clear away the term sheet "gotchas" such as liquidation preferences. Be skeptical about VCs who claim added value such as human resources, patent services, and all sorts of frills. These are the legacy of the late '90s stampede and, frankly, VCs are better at the money thing.

Nor is a start-up necessarily a job for life. The maximum life expectancy may be a year or less. But then, jobs in big companies are seldom for life, either! If you are part of senior management, don't mislead recruits on this point, either.

Be your own person. It is your business, so act responsibly. Experience and knowledge is invaluable when it comes to avoiding the traps (see above). If you don't have a particular facet of experience, don't fake it; find it. There's a lot of free(ish) help out there if you are prepared to network for it, and a variety of opportunities. Don't be shy to ask other entrepreneurs. Most are only too willing to help. Most VCs have their own networking events; these are a great opportunity to meet a whole variety of useful people. Companies such as Library House are acting as facilitators. And hey, if it doesn't help, someone has to drink the champagne!

And finally, finally... don't forget you have a life. You only have one shot at it. The start-up is very important, but so is your family!

4

Confessions of a Start-Up Junkie

Mike Redman
Entrepreneur and Consultant
United Kingdom

This chapter discusses the experiences gained from a predominantly industrial career and through a multitude of spin-outs with varying success. It emphasizes the need for people, trust, and respect within a small team and what happens when this all-important trust breaks down. Finally, some key topics to consider for a successful entrepreneurial process are offered.

4.1 Background

My first experience with a new business start-up came in January 1979. I had been working in the R&D laboratories of Xerox, where despite all our best efforts, the company's business had begun to be eroded by competition from Japan Inc. Although I had missed the first "headcount exposure," as Xerox euphemistically called it, I did not escape the second phase, and by Christmas 1978 I found myself out of work for the first time since I left school in 1962. (Why, oh why, do companies like to make staff redundant around Christmastime?)

However, before I had even banked my redundancy check from Xerox, I had received a job offer from Standard Telephones and Cables (STC), which at that time was part of the mighty International Telephone and Telegraph Corporation (ITT). I was to join a brand-new group, the New Product Development Unit (NPDU), which was being set up to provide a product-development service for the whole of STC. I duly reported to an empty warehouse in Harlow on the first working day of January 1979, along with several of my ex-Xerox colleagues. In fact, NPDU was, in its early days, staffed by many former Xerox personnel, probably because the head of the unit had previously been head of a division at Xerox!

My recollection of this start-up is that it all seemed very easy. We had a facilities manager who soon had all of the furnishings, fixtures, and fittings on order, and STC bankrolled the financing for the operation without too many difficult questions. I think we had a business plan, but the boss didn't speak about it too much.

Over the next four or five years, NPDU grew into a busy unit providing exactly what it said on the outside of the building: new-product development for STC. We worked on telephones, digital pagers, voice-messaging systems, speech recognition, a small private automatic branch exchange (PABX), and pretty much anything else that we were asked to do by the divisions of STC. A rather splendid product named Executel, which I suppose was a desktop forerunner of the personal digital assistant (PDA), was launched and won a Queen's Award for Industry. However, by year five things had started to go wrong for STC; the company had been sold off by ITT and was now having to stand on its own two feet financially. STC's major customer, the Post Office, was about to be privatized and become British Telecommunications (BT), and the cozy relationship between these two organizations was soon to end. To add further to its troubles, STC had lost its one-third share in the development of the UK's digital telephone exchange, System X. The writing was on the wall for STC and NPDU.

It was, however, decided that NPDU could reinvent itself as the Product Technology Centre (PTC), an organization that could sell its services to any company that was prepared to pay. All of the staff were packed off onto a week-long course on how to be good salespeople (run, you've guessed it, by an ex-Xerox sales trainer!). An up-market, multipage sales brochure was prepared, bearing our new name on the cover, together with all the accoutrements we needed to become rivals to PA Technology. Armed with these, the senior managers set off prospecting for clients. My company car, which up until now had just been a perk of the job, became vital to my everyday work; I would regularly drive more than 1,000 miles a week to make presentations to prospective customers. The unit had gained a good track record from the work it had done for STC, so we found it relatively easy to find buyers for our product-development skills.

However, conflict of interest soon raised its ugly head, and we were prevented from taking up some of the work that we found. For example, I had secured a development contract with a Swindon-based pager company. We were prevented from doing this work because STC was in the pager business and, although PTC was supposed to stand on its own two feet and be financially independent of its parent, STC would not allow us to work with companies it regarded as competitors. This rather limited what we were able to do, but nevertheless we were successful in gaining development work. According to our bookkeeper, we were well on the way to financial autonomy.

Unfortunately, big-company politics were also at play. Detractors within STC were vehemently opposed to the very existence of PTC, including the financial comptroller of our unit. He had already found himself another job within the company and was carefully massaging the figures to cast PTC in a bad financial light. Needless to say, the detractors won; and at the start of September 1985, it was announced that PTC was to be closed and the entire staff made redundant. This came as a severe blow to most of us; we had worked enthusiastically and professionally to build up a very successful product-development unit, and we were now to be rewarded by redundancy.

However, our new-found skills as salespeople were put to good use: by the time the closedown arrived at Christmas 1985, most of the staff had either secured new jobs or were well on the way to finding them. I had been given an extra two months' work after Christmas, overseeing the successful completion of a couple of development contracts, but I had nevertheless managed to obtain two offers, one of which was with Raychem in Swindon.

The fate of my erstwhile employer was sealed by now. Its share price had plummeted, and within a couple of years this once-great telecommunications company had disappeared without trace—its divisions absorbed into other companies or simply closed down.

I joined Raychem at the end of February 1986, in the Division for Utilities Ventures, or DUV for short. DUV had been formed to manufacture and sell products to the power utilities and was developing a system for installing optical fiber cables on overhead power lines. This would effectively enable a power utility to become a telecom supplier, provided the country's regulatory body would allow it. The system, known as Rayfos, used a specially designed cable and a machine that wrapped the cable around the overhead conductor. It included a whole range of fixtures and fittings, including polymeric insulators that were needed when the cable was installed on phase wires.

I remember only too clearly attending my first management meeting in DUV. We all sat around the table, with each manager giving a short presentation on what he had achieved in the last month. I was not expected to contribute, having only been with the company for a fortnight, but I listened carefully and didn't particularly like what I heard. The sales managers started the meeting and gave a summary of sales achieved and prospective sales. It became very clear that the sales revenue would not pay the salaries of those sitting around the table, let alone the cohorts out in the labs and production area. DUV appeared to be a vast empire without the sales to support its expensive ways. After the meeting, I voiced my concerns to some of my peers. I was told not to worry; Raychem would support the operation throughout its loss-making period of development and growth.

Later that year, I found my initial concerns were well founded. When I returned from holiday in the summer of 1986, I was informed that DUV was to be severely slimmed down and my job had disappeared completely. However, I was not as unfortunate as one or two others in the division; I was at least to remain employed by Raychem. For the next year, I found myself running the electronics group in Raychem's large Research and Development Division. This wasn't a particularly scintillating job, running a team of seven electronics engineers and an electrician who provided an electronic design-and-build service to the many groups within R&D. It did, however, pay a handsome salary and gave me the background knowledge in cable-making and cable-making machinery that I have been using for the past 20 years. I also became acquainted with Vibetek, a novel polymeric piezoelectric cable that was coming to the end of a successful development program. It had been developed primarily for the manufacture of hydrophone arrays for use in submarine detection systems. (The Cold War with the Soviet Union was still active and vast sums of money went to developing technology to fight a "hot war" that sane-minded folks hoped would never happen.) I had already identified other uses for Vibetek, including making pick-ups for musical instruments. I built a piano pick-up for Bob Hall, a blues and boogie-woogie artist who was Raychem's patent lawyer at the time. I also designed and built a prototype perimeter security system that used Vibetek as the sensor; this application has been close to my heart ever since.

My next opportunity to join a start-up came in the summer of 1987. Raychem, like most large companies, was a rumor foundry, and stories were circulating about the imminent sell-off of both the Rayfos and Vibetek businesses, but not necessarily to the same buyer. So it was, on a Friday afternoon in late July 1987, I sat in the office of Ray G., one of the Rayfos sales guys. He confirmed the rumors and, to my surprise, asked whether I would like to join the new operation. I took less than 50 microseconds to reply in the affirmative, and within two days I was on a plane with Ray G. and Rayfos' technical guru Jim B., traveling to Newcastle to meet and be interviewed by the joint managing directors of Cookson Group.

This meeting was a success; we all got key jobs in the new spin-off. By the autumn of 1987, a new limited company had been formed, which we decided to call FOCAS. FOCAS stood for Fiber Optic Cables, Accessories, and Sensors, to reflect the business interests of the new company. We didn't think the name was very clever, but at the time we couldn't think of anything more appropriate. One really exciting aspect of the new venture was that the 13 people who were giving up their secure, well-paid jobs at Raychem to join FOCAS were to be offered 20% of the share capital in the business, so we each acquired a significant number of penny shares in the company. The deal was that if we left FOCAS within five years, the shares had to be sold back to Cookson at face value; after this, they would be bought back at an agreed-upon independent valuation of the business—very effective "golden handcuffs," guaranteed to prevent key players from jumping ship in the formative years of the new business. What still never ceases to amaze me, when I think about this time, is the number of my former Raychem colleagues who told me how they thought that joining FOCAS was the worst career move I would ever make. How wrong they all were!

At the first management meeting, I was handed the job of finding and setting up accommodations for the new company. Although Raychem had agreed that FOCAS could remain on its premises for the immediate future, we needed to find our own factory, laboratory, and office accommodations as soon as possible. I found myself spending whole afternoons driving around the industrial estates of Swindon and its surroundings, looking for a factory unit or evaluating units with an estate agent. This, as anyone who has done it knows, is a pretty soul-destroying task. You never actually find the ideal unit, and I eventually concluded that the only way forward was to modify an existing unit.

However, this was not as easy as it should have been. My colleagues were singularly unhelpful regarding the specifications and requirements. They were unanimous in saying that the office and lab spaces must have windows, but that was about it. Floor space I could only guess at; everyone was suitably vague. Of course, we would all have liked the vast hangar-like space for production and the palatial offices that we had at Raychem, but this clearly wasn't going to be possible on a limited budget. And what about the budget for acquiring and setting up our new premises? We most certainly had a business plan, but it never included a statement delineating how much could be spent on the start-up. The whole operation was bankrolled by Cookson Group; but throughout the time I spent setting up the factory, I could never find out from our new managing director (MD) precisely what my budget was. The only feedback I ever got was to be told (by the MD) that I had overspent, but I was never told by how much. It wasn't millions, or even a hundred thousand, as the whole operation cost less than that. It would have been really useful to know what my budget had been set to; had I known this, I might have slept a little easier.

However, less than 12 months after the company had been formed, FOCAS Limited moved into Unit 4 Cheney Manor Industrial Estate, Swindon—a 10,000 sq. ft. warehouse that had been converted into a 12,000 sq. ft. manufacturing unit, complete with an R&D facility. This had been achieved by installing a large mezzanine floor and rebuilding one of the walls so that we had windows on two stories.

The production equipment was moved in, and FOCAS was ready to start manufacturing fiber optic cables in its own factory. For a first effort at setting up a factory, I don't think I made too bad a job of it. I even overheard the MD of Cookson Group congratulating the MD of FOCAS, at the grand opening of the new building, for what he had achieved for such a low spend!

It is pretty obvious that any new business will encounter problems during its early years, and FOCAS was no exception. The jacket we had selected for the wrap-on fiber optic cable, now called Skywrap, was a special polysiloxane rubber with a rusty red color. It had all the properties required of it, with an additional one that neither we nor the compound manufacturer had anticipated: in some parts of the world, the birds mistook the wrapped cable on the overhead conductor for a giant worm! Soon, flocks of birds had completely destroyed kilometer lengths of Skywrap. This caused a major panic in FOCAS; we had to suspend all manufacturing and installation while a new material was found and qualified. I think it is true to say that without the financial support of Cookson Group, the company would have gone under during this critical time. No business plan could have foreseen the impact our feathered friends had on our business. However, we eventually settled on a cross-linked polyolefin material for a jacket—this time in black—and the business was soon back on track.

As the business grew, we took on more staff, opened a sales and manufacturing office in the U.S., and introduced more products into the portfolio that we were successfully selling to power utilities worldwide. We moved out of the original factory into a unit of 50,000 sq. ft. in an adjacent industrial park. I was not involved in its set-up, but I was surprised to find that somehow we had managed to fill such a large building overnight! Even though all the managers were shareholders in the business, internal politics and management rivalry became the norm, and the MD survived at least two attempts (that I knew of) to usurp him through a boardroom coup. However, the company prospered, and its turnover soon rose to well in excess of the $10-million level that Raychem had confidently stated and our original business plan had predicted the business would never reach. What hadn't been predicted was the mix of products that would achieve this level of sales; we were also manufacturing optical ground wire and datacom cables. There were still problems, of course, not least of which was a lengthy and expensive court case with a competitor, but times were good. Deregulation led to growth for the telecom business worldwide. FOCAS secured a significant part of the contract for supplying the fiber-optic cable to the UK National Grid's telecom spin-off company Energis. This contract kept the Swindon factory very busy for nearly two years. It also ensured that when we sold our stake in FOCAS back to Cookson group in years six and seven, we made a very handsome profit on our 1p shares.

Although FOCAS flourished, the fortunes of its parent, Cookson Group, started a roller-coaster ride of ups and downs, with the net trend being down. The shares I had bought in Cookson for more than £8.00 each were now worth around £1.50. At FOCAS, we were all being exhorted to economize in everything we did, except of course for the privileged few who were still winging their way across the world in the class to which they had become accustomed. I was still a check signatory, and I recall being outraged when asked to sign a check to a travel agent for well over £20,000 to cover one month's airline and hotel bills for three or four people in the company.

By the time I got my third new-business opportunity, early in 1997, I thought I was ready for it. After all, I'd had a fair bit of practice over the years, so this time it was going to be a breeze. The decision was more or less thrust upon me when I was made redundant again. Cookson Group was planning to sell the FOCAS business lock, stock, and barrel to AFL in the U.S. and clearly wanted to make it a more attractive proposition by getting rid of the more expensive members of the company to slim down the payroll. The MD who had been with us from the start had already left, just before Christmas 1996. I was in the group that followed.

FOCAS provided the 30 or so of us on the redundancy list with an "outplacement counselor." Curiosity took me along to the first meeting with this lady. She started the interview by asking me how I felt about being made redundant by FOCAS! My answer (unprintable here) made her tell me that I really shouldn't feel any anger or bitterness towards the new management, despite having devoted nearly ten years of my life to the company. After a few platitudes, the final straw came when she told me that, at 52, I shouldn't expect to get a new job very quickly, it was going to take "quite some time."

Part of the slimming-down operation at FOCAS involved dropping the manufacture of Vibetek, the range of novel piezoelectric cables that had originally been part of the Raychem spin-off. Sales of Vibetek had, by 1996, become a minute part of FOCAS' multimillion-pound turnover and the product had inevitably been sidelined. I felt there was still a lot of life left in this product, particularly if some new cables could be added to the range. During the previous year, I had also become involved in a very interesting technology development, HYDROFAST, working with the University of Strathclyde and MGS Geosense. The aim was to develop optical fiber water and pH sensor cables, using a range of novel hydrogel materials that were under development at Strathclyde. FOCAS was receiving about 20% of the costs incurred by this project through a DTI LINK grant. However, the new management did not consider this project to be worth pursuing, so it was immediately stopped.

I approached the new MD and inquired whether a deal could be struck to take Vibetek and HYDROFAST to a new company, to be started by David A. (another soon-to-be-redundant member of FOCAS' Technical Group) and me. This was met with a fairly enthusiastic response, and a price was duly provided for supplying the Vibetek business, some of the manufacturing equipment, and the stock. I suppose it shouldn't have been a surprise, but there was a huge gulf between FOCAS' price tag and what I considered the business's worth. There didn't seem to be any room for negotiation, so I didn't try. I just went about the rather depressing business of tying up the loose ends of my old job and clearing my office.

About 10 days before my last day at the company, I received a visit from FOCAS' accountant, who inquired how the purchase of the Vibetek business was proceeding. I told him that it wasn't because the price was far too high and the new MD didn't appear to want to negotiate. He asked what I was prepared to pay for the business; I told him, and he went away saying that he would look into it. Within a couple of days the deal was done and dusted, and the FOCAS solicitor was busy drawing up an agreement to transfer the assets of the Vibetek business to our new, as-yet-nameless company.

The advice here is that if you are trying to acquire part of the business of your former employer, don't go into negotiations too early in the process. Be noncommittal and act disinterested. They probably want to dispose of it far more desperately than you actually want to buy it. It will give you valuable time to investigate the business and put together a business plan; due diligence is much easier from inside the vendor's camp!

On the subject of business plans, I am not particularly enthusiastic about spending vast amounts of time putting together a highly detailed one. Yes, I have done this, but in practice they never seem to work out as planned. For example, the business plan for FOCAS was highly detailed, and in the end we achieved the projected sales targets, but not by selling the products that were originally proposed. A number of programs are available for speeding up the assembly of a business plan; it good to have something to show the bank manager when he asks, but he will probably only be interested in cash-flow projections and whether the company will require a business loan from day one. Some sort of a mission statement is vital; this needs to state what you are going to do and the company's aims and aspirations.

As far as a detailed financial plan is concerned, just concentrate on the cash flow. It is vital to know how the bills are going to get paid and to have some sort of worst-case contingency plan. Suppliers are pretty unsympathetic during the early days of a start-up; you probably will not get the 30-day credit that you will once the company has established a trading record. All the bright ideas in the world will come to naught if you aren't selling as much as you expected to and the landlord is banging at the door for the rent. My advice is to have a contingency fund for these crisis moments; otherwise, the company is not going to get much past "GO." Oh, and resist the temptation to go out and lease or buy a shiny new BMW; driving a flashy car may make you feel like a managing director, but the money in the early days is better reserved for more important purchases.

Having agreed to buy the Vibetek business and some of the manufacturing equipment, setting up a company and finding premises became a priority. Choosing the company name, Ormal, wasn't particularly easy. We wanted the name to reflect the optical-fiber sensors, such as HYDROFAST, that we were planning to develop. However, a quick search revealed that names containing "opto," "senso," and the like were already heavily used, so I was forced to look into more obscure, possibly made-up, words. Eventually, JRR Tolkien came to the rescue with "Ormal." I found this in the index of Tolkien's Silmarillion: Ormal and Illuin, according to Tolkien, were two great lamps that illuminated Middle Earth before the glory days of Gandalf, Bilbo, and Peter Jackson. Unbelievably, a search at Companies House revealed that there already was one very small company in existence using the name Ormal, but as it was in the financial sector there wasn't likely to be a conflict. So, we went ahead with registering Ormal Limited as our company name.

Registering new companies was something I hadn't done before, so I engaged a local solicitor who claimed some knowledge of company law and asked him to do it. I was naïve enough to think he would set up the company for us by dealing directly with Companies House. Instead, he did what we could have done in the first place, bought a ready-made company from an agency that specializes in setting up companies and then changed the name to Ormal Limited. For this, we got a bill that was not insignificant, considering the small amount of work that was actually involved. Nowadays, Companies House has a strong presence on the Web (http://www.companieshouse.gov.uk) and will provide all you need to know about setting up a new company or changing the name of one bought off the shelf.

Finding a suitable factory wasn't too difficult, although it was expensive. Our contacts at the University of Strathclyde had suggested approaching the Scottish Development Agency (SDA) with a view to setting up the whole operation over the border. This was rapidly discounted when we were told by a gentleman at the SDA that he couldn't even let us see a list of factory units that were available in the Strathclyde Region. By contrast, we had received comprehensive lists of accommodation and offers of further assistance from, for example, the Welsh Development Agency and, nearer to home, the West Oxfordshire District Council.

In the end, we decided to locate our new business in the town we had grown to love during the previous 10 years, namely Swindon. This eliminated the need to move home and, more important, provided us with a pool of labor skilled in the art of wire and cable manufacture. So, we started to look around Swindon and soon found a unit of about 290 sq. m. on a fairly new industrial estate in the town. The unit was clean, devoid of any interior fittings whatsoever, and ready for immediate occupation, so we gave the details to our solicitor and asked him to get things moving, because any day we would become the proud owners of the Vibetek business together with a significant amount of the equipment needed to produce this innovative piezoelectric cable.

As I had found with the FOCAS start-up, taking a lease on a factory unit is an eye opener; with Ormal, I found that when the unit is owned by a large organization like an insurance company, acquiring the lease is decidedly stressful. Our solicitor helpfully told us that the length of a lease is usually in direct proportion to the size of the organization letting the building, so our lease was very, very long. It contained, for example, an interesting clause that prohibited, should the landlord decide to enforce it, the use of mobile communications, both wireless and optical. Even when it was pointed out that nearly everybody had a mobile phone in 1997, they insisted that this clause remain. More problematic was the requirement for a rent guarantee, something that we hadn't budgeted for in our business plan. We were told that we would either have to put up a deposit of six months' rent and service charge, including value-added tax (VAT) for the life of the lease, or hand over the deeds to our homes. The solicitor strongly advised against the latter, so we were left contemplating the handover of more than £10,000 of our much-needed capital for an indefinite period. Very generously, our would-be landlords assured us that we would get the interest on the rent deposit; what they didn't tell us was that they were going to deposit it in a standard bank account paying the lowest interest rate going. By this time, it seemed as if a queue of people were standing in line to take money out of our embryonic company, without providing much in return. We had handed over about £12,000 of our own hard-earned capital before we were eventually given the keys to our factory unit and were free to properly begin our new business.

The factory, as mentioned, was an empty shell. The first meeting of the HYDROFAST LINK team was held over a picnic lunch in the grounds of the industrial estate. In addition to not having been able to take delivery of any furniture, on the day we received the keys, we discovered that there wasn't a single 13A outlet into which could be plugged the new company kettle—it took an emergency visit from our electrician to install the tea-making facility! A great deal must be done to convert a factory shell into a fully functioning unit with office space, telephones, laboratory facilities, and, most important, the factory area. Whatever grandiose plans we had for future technology development, we needed to make and sell the existing products to fund these plans. I was relieved to discover that my previous experience setting up the first production unit at

FOCAS served me well. I knew exactly what needed to be done; the only difference was that I was now spending my own money, not somebody else's!

We soon had an electrician putting in the lighting and three-phase distribution for the production machinery, and a contractor erected an office area at the front of the new building; it all looked very businesslike. While David organized the production and quality control (QC) area, I set about talking to the existing Vibetek customers and trying to find new ones. We were given invaluable practical assistance by one or two of our friends at FOCAS, particularly Kevin P. and Mike Foote. Mike F. eventually joined us full time at the company and remained a loyal friend and employee until his untimely death in 2004.

I had also been in discussions with Ray G., the erstwhile managing director of FOCAS. He was anxious to invest some of his money in the new company, and we had agreed that he would join the board of directors as chairman. His role in the company was as yet undefined, but he did agree that he would come in on a regular basis, roll his sleeves up, and help with whatever needed to be done. After a couple of visits, it become obvious that Ray was not really prepared to provide any sort of manual assistance and that his interest in the company was purely financial; in a nutshell, he wanted to get rollover relief on the capital gains tax that he had paid when he, and indeed the rest of us, sold our FOCAS shares to Cookson Group. What finally clinched it was a telephone call from Ray late one Friday afternoon, demanding that I issue the shares in the new company straight away. His accountant had, he claimed, told him that if the shares were not issued immediately, we would not be able to claim rollover relief on our new investment. I merely pointed out that as yet he hadn't provided any financing for the company, the whole operation had been bankrolled by my wife, David A., and me, so I didn't think that he or his accountant was in a position to demand anything.

This appeared to be the wrong thing to say. He told me that he didn't think investing in our company was going to be the right thing for him to do, and he wished to pull out completely. Clearly, this wasn't going to be an insurmountable problem for us; Ray G. hadn't provided much in the way of practical assistance, and all we owed him was the expenses on his car for a couple of trips that he had made with us. It did, however, mean that my wife, somewhat reluctantly, had to become a major shareholder in the company; she had been lending the company working capital, which should have been provided by Ray G., during the start-up. With hindsight, this pull-out was a blessing in disguise; there was a huge gulf in management style, culture, sense of humor—and just about everything else between Ray and me—which would inevitably have led to conflict later on. So it was that my wife cheerfully provided us with much-needed assistance on the sales and office administration front and kept the books with great accuracy for the first eight years of the company's life. I am also pleased to report that Ray G.'s accountant was wrong about the rollover relief; we did eventually receive a very useful check from the Inland Revenue.

In due course, the Vibetek manufacturing line was set up, and it was with considerable pride that I quality-tested and passed the first product that emerged. This was a very welcome event; I had gained a significant export order that needed to be shipped in the first two weeks of September 1997. While we were setting up the factory, I had been steadily selling the stock we had bought from FOCAS, so pride was not the only emotion I felt when the first cable came off the production line. I also felt great relief.

Looking to the future, we had been trying to continue the HYDROFAST program, and I had been busy filling out forms for the DTI to apply for the LINK grant. This was fraught with problems, not the least of which was that David A. and I had decided not to draw a salary from the company during the first year of its operation. We didn't need to; FOCAS had paid us off quite generously. In fact, I received more from the company when I left it than I would have if I'd remained employed with it for another year. This absence of salary conflicted with the requirements of the DTI application; the grant couldn't provide any payment for your labor if you weren't actually paying yourself anything. I did question whether we could pay ourselves generously for the days that we were working on the HYDROFAST program but was told that the grant rules prohibited this; salary couldn't be varied over the year, despite this being what most company directors seem to be advised by their accountants to do. So we didn't actually get very much financial help out of the DTI for HYDROFAST during the first year, despite having to spend a fair amount of time actually working on the technology and attending meetings in Glasgow. My advice to any small company thinking of entering into one of these collaborative schemes would be to look very carefully into what you are going to have to put into the scheme as well as what you stand to get out.

HYDROFAST is an amazingly clever technology, but in nearly 10 years with Ormal, we still have not been able to find a commercial outlet for it. Towards the end of the HYDROFAST project in 1999, the company developed a very simple (and inexpensive) copper-based water-sensor cable, which we thought would be a best seller in the residential market. However, the Aqualert Water Leak Alarm, even at a retail price of £15.00, was never a big seller. It even had the dubious honor of appearing as a prize in a tabloid newspaper "advertorial" competition, featured next to another competition that was offering a sex aid as a prize! I am also amused to see that an Aqualert occasionally turns up on eBay! All of this development effort was not entirely wasted, however; the technology is used in a sensor that Ormal builds for an original equipment manufacturer (OEM) in the business of environmental monitoring in offices, banks, and computer rooms. These units do sell in significant numbers.

We completed our first year with a fairly healthy turnover of nearly £100,000, made a very small profit, established the company as a manufacturing operation, and put in place all of the necessary quality-assurance procedures. Moreover, the HYDROFAST project was taking shape, and David and I were looking forward to expanding the business during its second year, with the help of another FOCAS colleague, Robin D. Robin had expressed an interest in joining us from the start-up in 1997, but he had a young family to support and we had agreed that the risks were too high to leave a well-paid job if he didn't have to. By the end of March 1998, the outlook seemed reasonably rosy, and Robin D. left FOCAS and joined Ormal as a 10% shareholder and director of sales and marketing. With hindsight, taking on Robin D. was the worst thing that I did in my entire time in the company. It wasn't just a serious mistake; it was a near fatal mistake.

David and I always intended to add a cable strander to our equipment as soon as the business became established, so we could manufacture multiconductor or multifiber cables with a higher level of complexity than the simple coaxial construction of Vibetek. David had even looked at a second-hand cabler. This all changed with the arrival of Robin D.; he told us that the market for this type of cable was very competitive and persuaded us that we should take on an agency to sell cables manufactured by a Korean company with whom he had developed a good relationship while at FOCAS. While this didn't fit with our original strategy, we could see merit in the idea and allowed Robin D. to proceed. At my first meeting with the Korean company's representative, I was surprised to be told about what I now believe is a fairly common business practice with some of the larger Far Eastern companies. Our commission on cable sales was agreed at 5%, of which 4% would stay within our company; we would deposit the remaining 1% in the private bank account of our Korean account manager.

At first this agreement seemed academic; Robin D. turned out not to be a particularly good salesman. During his first year with the company, he didn't sell any cable from our Korean supplier or find any new business for our existing products. The accounts for our second year of trading showed that our sales were significantly lower than in our first year of business and that we made a net loss. To add insult to injury, I had to inject more cash into the company as a loan to provide the cash flow to pay the wage bill.

During year three, things started to improve. Robin D. had spent virtually his entire time pursuing orders for Korean cable and was starting to see some success. I am not entirely sure whether this success was due to his improving sales technique or simply that it was the start of the telecom bonanza that occurred around this time; product shortages were occurring, making it much easier to sell product from the Far East. Robin D. had even started to sell telecom-grade optical fiber, supplied by our Korean friends, to a couple of cable-making companies. The ethics seemed questionable; we were selling fiber to British cable-making companies with whom we were also competing in the cable marketplace. Our sales director was always able to reassure me that this was all right; I didn't need to worry about it.

We made some progress with HYDROFAST; I think by this time we had successfully demonstrated a pH sensor cable. Vibetek sales were steady, and we had sold a couple hundred sensor assemblies for traffic monitoring. By the end of year three, we were starting to see overall improvements in business performance; by now we had even started to pay my long-suffering wife for her monumental efforts in keeping the books, answering the phone, and organizing several mailshots singlehandedly.

Year four was a bumper year. The fiber-optic cable business went into overdrive; companies just couldn't get enough of our Korean-manufactured cable. Selling it wasn't really the issue; for most of the year we were rationing it. It is true to say that if we could have obtained more cable, we could have sold it. Likewise with optical fiber; we could easily have sold more than was made available to us. The company prospered, and so did our Korean friend. Every month we paid into his bank account more than we were paying ourselves, and we certainly weren't complaining about our pay. As I saw the year-end profit prediction grow, I started to discuss with my two co-directors how we should invest this money. Clearly, we had to retain a good bit of it in the company, but there was going to be enough to pay ourselves a decent dividend. I also could see that there was an opportunity to invest in a factory owned by the company, or possibly the company pension fund. David A. was always receptive to these suggestions, but Robin D. only seemed interested in the dividend. He went as far as to suggest that all the profit should be paid out as dividend and split in equal thirds among the three directors. Greed was starting to take over from common sense and fairness. Greed I could handle, but what I had not bargained for was the downright dishonesty of this person who at one time claimed to be my friend.

By the time we had filed year four accounts, which did show the expected big profit, there was a definite change in Robin D.'s attitude towards the business. We did pay a significant dividend, to the holders of three "A" shares, intended to minimize the tax we would have to pay, hastily issued to the wives of Robin D. and me as well as to David A., who was single. Had we decided to pay the dividend the way most companies would have, 70% would have gone to me and my wife, which didn't seem entirely fair to the other two. This first dividend payment was received without much enthusiasm by Robin D., and he immediately began asking when the next payment would be made. In the first few months of year five, it became apparent that sales of Korean-made fiber optic cable had taken a nosedive; I was certainly expecting to see an end to the telecom industry mini-boom, but not quite as rapidly as this. To cut a long (and painful) story short, our sales director had formed his own company and was using it to book the Korean cable orders, while still ostensibly working for Ormal and drawing his salary. He did this for seven months before we fully realized what was going on; my clever wife's detective work, with the help of a forensic computer company, unearthed the details of the plot. Robin D. finally handed in an unsigned letter of resignation, which told me that he was leaving Ormal and taking "his" cable business with him. I hired our local lawyer to sort matters out, which turned out to be a seriously bad move. He charged us a lot of money for doing little more than tell us what we couldn't do; by the end of year five, Robin D.'s company had taken complete control of the Korean cable business, and Robin himself had started proceedings in the Industrial Tribunal for unfair dismissal and a one-third share of Ormal's year-four profit, claiming it was "an unpaid bonus."

Although this action was eventually dismissed as a spurious claim, in fighting the case we spent a significant amount of money on legal and accounting fees. And, of course, during the eight months that we spent resolving these problems, the business suffered, not only through the loss of fiber-optic cable sales but also because David A. and I had not been giving our remaining business the attention that it deserved.

However, Ormal has survived. The company still manufactures Vibetek sensor cables and water-detection systems and for the past few years has been successfully moving into electronic product development for OEMs. I retired from the company when I reached my 60th birthday, something I had always planned to do, but I still do consultancy work for Ormal and others. Who knows? I may even start up again; I still have a few half-decent business ideas!

4.2 Key Topics for an Entrepreneur

Over the years, my experiences in new company start-ups have taught me a great deal, but I also realize that I have made far too many mistakes. I am most definitely not a start-up guru. But finally, for what they are worth, I have summarized a few pointers for anyone contemplating taking the bold step forward. These bullet points are not in any particular order of importance.

- When setting up a company, have a shareholders' agreement drawn up. This can be a fairly basic document, but it needs to set out clearly such things as what happens if a shareholder wants to sell his shares or to buy more shares in the company.
- Directors' loans to the company also need to be the subject of a written agreement; otherwise, you may find that you have lent significant amounts of money to the company without any record of the interest rate that is to be paid and when the loan is to be paid off.
- All employees, and particularly the directors, of the company need to have employment contracts that clearly set out what their responsibilities to the company are and vice versa. This probably won't stop a potentially dishonest director who believes it is his right to run off with half the business, but it may help you rectify the situation if he does.
- When setting up a business with friends, don't assume that they will remain your friends. Experience shows that once the profits start to roll in, friendship and fairness may play second fiddle to avarice and dishonesty.
- If you need legal advice, don't be tempted to use the local solicitor who did your house conveyance and wrote your will. Most simply do not have the necessary in-depth experience of business law. Go to a legal practice that specializes in corporate law.
- Make a business plan, but keep it short and simple. Start with a mission statement and work from that. The key to success is understanding what you are going to make and sell, with a realistic outline of what the financial figures will look like. Cash-flow predictions are critical to

success, and they must have adequate contingency for the unexpected built in. Treat the business plan as a working document; be pragmatic and don't be afraid to revise it every so often as you see the way the business is going. You will obviously have to be more creative and provide a lot more detail in the business plan if you are trying to raise a lot of venture capital.

- Directors should be modest in their transport aspirations when the company is young; a Rolls Royce may look good in the car park, but it won't help pay the rent or workers' salaries. Similarly, be modest when furnishing the premises; a lot of fairly cheap (but excellent quality) second-hand office furniture can be had, mainly thanks to companies that have failed because they got the cash flow wrong!
- Join your local Chamber of Trade and Commerce. You may think as the owner of a high-technology company that you don't need to network with anybody outside your immediate sphere of interest, but sometimes this can bring very positive benefits. Discussing a problem with a noncompetitor can often be very helpful; they may well have been there, too, and have already found the optimum solution.
- When coming up with ideas for new products, it is always worth building a mock-up or model of your idea. If you can make it work, so much the better. For example, a working emulation of a product on a PC will greatly help you sell the concept to business angels, banks, and prospective customers.
- Always try to get patent protection for new ideas, register trademarks, etc. Unfortunately, this can be an expensive process, but the initial filings are relatively cheap.
- When setting up your Web site, register the alternative URLs if you can. Initially, we only registered http://www.ormal.co.uk, but within six months somebody called me trying to sell us http://www.ormal.com at a grossly inflated price. When I refused to buy it, its owner parked a gambling site on it! We had to wait two years before we were able to buy www.ormal.com at a reasonable price.
- Market research is vital when deciding whether to proceed with the development of anything new. Properly carried out, market research will enable you to make a rough estimate of the sales levels of your product and will provide valuable insights into the necessary features. Overlook this at your peril. For example, I think it doubtful that Sir Clive Sinclair would have proceeded with the development of his ill-fated C5 vehicle if the concept had been made the subject of a market research program. There are, of course, notable exceptions for which market research would not have predicted a product's runaway success; Xerox' first photocopier is a classic example. Chester Carlson, the inventor of xerography, had extreme difficulty finding financial backers to develop his idea; nobody believed the concept would gain such widespread acceptance in the

office and factory. Carbon paper was perfect for taking copies of typed sheets!

- Every new product idea requires a product champion, somebody who believes in it enough to fight for its very existence in the early days of development. I would venture to suggest that without a product champion, any new product, no matter how good, is doomed to failure, or at least to mediocrity.
- Selling is not something that comes naturally to most people, but most people can be trained to sell. I would strongly recommend that anybody starting a new venture obtain some sort of sales and marketing training; it is so much easier to obtain a first appointment with a prospective customer if you know how to go about it and you are confident that your sales brochures are up to standard. Even in these days of Internet shopping, you still must know what features a Web site needs to be an effective sales tool and how to drive the right sort of traffic to it. Most of the techniques required can be learned from cost-effective short courses provided by organizations such as Business Link and local Chambers of Trade. After that, practice makes perfect!
- A successful product only needs to be about 95% right, provided it is launched at the right time and at the right price. An absolutely 100% perfect product will fail if it is too expensive or too late; it will always be beaten by the competition.
- Whatever you are trying to sell, it is crucial to establish the correct selling price from the day the product is launched. Buyers usually have a perception of the maximum price that they are prepared to pay for an item; provided that you stay below this perceived value, the sales volume will not be adversely affected. Increasing the price of a product does not necessarily bring about a significant reduction in the number of products that are sold. Conversely, if you halve the price of your product, you won't necessarily double the sales volume, but you will certainly reduce your profit.

5

Being an "Intrapreneur" and an Entrepreneur in the Optoelectronics Industry

Michael S. Lebby
President and CEO
Optoelectronics Industry Development Association (OIDA)
United States

The author discusses his experiences as both as an intrapreneur at Motorola and Intel and as an entrepreneur as the CEO of Ignis Optics. Intrapreneurship comes from founding projects, programs, and new technological directions within a corporation, while entrepreneurship is the process of founding technological programs in standalone companies. Early in the author's career, he realized that there was reason to put long-term strategic goals in place; so that as opportunities arose, they could be evaluated from both an intrapreneurial and an entrepreneurial standpoint. The author discusses the virtues of knowing the strategy up front and worrying less about the tactics, but making sure the tactics align with the strategy as experience grows.

5.1 The Beginning

The will to succeed began at a very young age, perhaps in the single digits, driven by the natural desire to be first in something, whether in line for school meals, first on the bus, or first in mathematics. This drive has been a friend during my career and has had a profound affect in the level of my success. At "infant school," I wanted to become a soccer star or an astronaut, like most of my peers at that age. As years progressed to high school, those ambitions led me to become someone who is very technical and technological but also understands the business aspects of the technology. In the UK, a person with degree potential is often asked by high school teachers what they may want to read at university (NB: in the U.S., specialty focus seems to occur at the master's level, while in the UK, it is at the bachelor's level). At the age of 10 or 11, I somehow knew that I would pursue a technical degree, most probably a mix of physics and mathematics, and then later become more business-oriented. During lunch period at age 13, I can remember telling a teacher that perhaps I should be an engineer or scientist until the age of 40 and then be in a position to lead a science or engineering team. The teacher not only concurred but nearly choked on his food!

I remember thinking at the time that if I was reasonably successful, I would be promoted to a position where I would do less science and move toward undertaking business and leadership roles. Interestingly, that level of thought in the early '70s accurately reflects my career today.

5.2 The Planning

Planning an entrepreneurial career did not begin in high school. In high school, the strategic direction was advised by my teachers, and I chose to follow that direction, but I really didn't know the best choices to make for the classes. However, deep down I was confident in the strategic direction, so I developed the attitude that I would evaluate and, if appropriate, pursue every opportunity that came my way.

Most degree candidates in the UK matriculate from high school to a college or university. I chose an alternate route simply because the opportunity appeared. While I was in high school, a local government facility offered eight apprenticeships in electronics and telecommunications. The apprenticeships consisted of a four-year training program in conjunction with part-time college studies, which led to craftsman or journeyman status at the technician level. I remember thinking that surely the best route to becoming a scientist or engineer would be to take high school graduation courses for entry into university. Looking at published university prospectuses, I noticed no courses in electronics. I was quite disappointed to realize that the subject I wanted to study was not readily available in the UK university system.

However, I noticed that these courses were offered by polytechnic colleges and wondered about the best route to get into these institutions. I learned that polytechnic students fell into two categories: those who did not achieve high enough grades for university or those who were supported by their company to pursue technical training. Because the route to engineer or scientist was much more difficult and in most colleges nearly impossible, I chose to apply for one of the eight apprenticeship seats—along with 300 other applicants! This was a risky option, but I believed that a strong skill base in electronics would be beneficial. I was fortunate enough to be accepted and decided to leave school and head into industry at the age of 16.

Part-time college study for technician qualifications was a part of the apprenticeship. I did well at college but only later realized the critical role that my experience in mechanical, electronic, optical, and telecommunications engineering would play in becoming a successful scientist post-doctorate. I also realized that a person excelling in the technical field both at work and college should probably opt for the direct-to-university route in the first place. After three years of technical training at college and formal apprenticeship training, my employer put my name forward for a national government scholarship competition for university. I understood that specialized courses in electronics were only part of becoming an engineer; it is the formal education process and discipline that is of real value. I succeeded at the national level interviews and was awarded a scholarship to any university in the UK.

The chance to move my career back onto my strategic plan was paramount, and I chose a down-to-earth engineering university in Bradford, Yorkshire, known for its strong telecommunications research and electronics engineering department. My plan was not only to use the skills learned during my apprenticeship but to add knowledge of design and theory. During my apprenticeship, I was fault diagnosing a UHF telecommunications transmitter/receiver using various electronic test and measurement equipment. I recall asking who designs the equipment if we were the ones to fix, fault find, and repair it. The answer was simply "those folks who have been to university to learn the theory." That answer was enough for me to push for the next step—university.

5.3 The Training

University was both simple and difficult. The few students who entered from industry excelled in all the practical experiments and laboratory studies, but suffered in academic classes, as their theoretical training lacked the rigors of the A-level high school courses. After playing catch-up for two years, I knew I was on par with the A-level students and could then think about how to focus on my strategy to become an engineer or scientist.

The particular university course that I had was called a "thin sandwich" course at the time. These courses were split evenly between industry and university training. Six months of each of the first three years were spent in industry and six months at the university. University work comprised the entire last year, to complete final examinations.

The industry training immediately following theoretical coursework is a wonderful concept that seems to be often overlooked in education systems today. In my particular circumstance, it allowed me to practice electronics design in industry after learning the concepts in the classroom. An undergraduate thesis in the final year involved computer-aided testing and measurement of semiconductor devices. This allowed me to specialize still further in semiconductor technology.

On graduation, I was offered a full-time position in the government research laboratories in RSRE Malvern (now Qinetiq) and a scholarship for a technology PhD at the university. The industrial research position was not only in a field that was directly on target for my strategy, but it offered real income, a huge attraction for a student. The alternative was a chance to pursue a graduate degree that required fulfilling both a technical PhD and a master's of business administration (MBA) in three years. This was a difficult decision because both alternatives offered tactical routes that aligned with my strategy. I chose the technology PhD and worked the first year on an MBA program at the Bradford Management Center. After the first year at management school, I was accepted as a graduate student at AT&T Bell Laboratories in New Jersey, U.S., and spent the next two years designing, fabricating, testing, and evaluating gallium arsenide and indium phosphide (GaAs and InP) electronic and optoelectronic semiconductor devices in the laboratory.

A number of published papers later, I was awarded both the MBA and PhD from Bradford University. The exposure to and excitement about learning the business aspects of technology were so great that, at the finish of the MBA, I majored in small-business planning. It was the MBA that made me realize that being either an intrapreneur or entrepreneur was not far away. Both roles were within reach, given the correct components and tactics.

During the MBA, I focused on learning the components that made entrepreneurs successful and completed a project on the subject as a prerequisite to my MBA thesis, a business plan to start a company. The results of the project were not completely satisfying, but they did teach me one thing: intrapreneurs and entrepreneurs are successful because of one simple trait—hunger. They are the hungriest, they want to survive, they want to win, and they have the will and drive to succeed, even at personal sacrifice. The question I had to ask myself was whether or not I fit the mold. Did I have this internal fire?

5.4 Intrapreneurship

After finishing the MBA, I briefly considered throwing in the towel on my PhD and getting a job on Wall Street earning a high salary. I didn't. It was probably one of the best entrepreneurial decisions I ever made. I believed that, armed with an MBA and a PhD, I would be able to differentiate myself among the thousands of other PhDs looking for research positions. As it happened, after two years of research at AT&T Bell Laboratories, the perceived differentiation of my MBA had expired because the research work did not use the business skills I had learned. Nonetheless, I continued the research path knowing that, at 27, I had both the PhD and MBA in my "hip pocket."

Accepting a position at Motorola in the corporate research laboratories, I started in earnest to use my research training from Bell Labs and the PhD program. During this period, I began to realize how an intrapreneur becomes successful inside a company. I put my energy not only into completing research tasks set by management but also into starting and driving new programs within the research laboratories. One program, OPTOBUS, grew to be staffed by more than 150 employees and had orders on the books for $90 million, with bookings for another $60 million. OPTOBUS was an optoelectronic interconnect with parallel modules, parallel fiber ribbons, the first high-volume production vertical-cavity surface-emitting lasers (VCSELs), and low-power bipolar complementary metal-oxide semiconductor (BiCMOS) array circuitry.

Motorola initiated production of OPTOBUS after four years of development that a colleague and I began in the research laboratories. The taste of intrapreneurship was sweet, but it was also sour. By the time the first big purchase order had hit the books, management had decided to close the program and reassign employees to other tasks. Four years in development took me from a practicing principal investigator in research with no subordinates to a section manager with up to 30 people on the project. The most frustrating part about the whole project was believing in its success even if others couldn't see the forest for the trees. This certainly was the case when Motorola's leadership decided to close the project down even with orders on the books. The lesson learned is that even if the corporate environment supports your intrapreneurial talents, even if you are allowed to file more than 300 patentable ideas (currently I have more than 175 issued USPTO utility patents and more than 50 international patents), the company still controls whether the project succeeds or dies, irrespective of the hunger of the intrapreneur.

5.5 Waiting for the Right Opportunity for Entrepreneurship

The intrapreneurial experience with OPTOBUS was critical to staying on course for my long-term goal: to be in a leadership role in technology by the age of 40. I was 36 when I departed Motorola and chose to work for one of the top three optoelectronics companies in the datacom segment. The three candidates were HP, Siemens, and AMP. HP at the time was suffering from internal political problems between their UK and U.S. divisions, Siemens was not executing its technology efforts well, and AMP was rebuilding its team because it was in third position and wanted to become leaders in this market segment. I chose AMP, which was acquired by Tyco International six months later; the division fell apart because of poor leadership. Clearly, my strategy to pursue more intrapreneurial projects was not working.

In AMP, I served as a business-development specialist, focusing not on the running of technical programs but rather on business deals, partnerships, and collaborations, and generally looking for business-based ways to help existing technical teams finish their tasks. After six months of Tyco leadership, I realized that this strategy was not going to work and considered my next move. It was 1999 and the optoelectronics bubble was in full swing. Finding a position at any company was relatively easy as most companies were expanding as fast as possible.

I joined Intel as a corporate investor. Intel wanted to find out more about this vibrant optoelectronics business and make some private-equity investments. The company needed someone on staff who would advise and guide them from a technical standpoint. I took the position and felt hungrier than I have ever felt in my life. Deals were pouring in from all over as our six-person opto investment team completed more than 30 deals in 18 months and, with the IPO events by New Focus and Bookham, even achieved a positive internal rate of return (IRR). I knew this position would bring me up to speed on the business aspects of technology, and also that if I completed a similar number of deals as I had filed patents, I would see just about any type of deal and negotiate any type of conditions that anyone would expect in this role. Additionally, I knew that if I could complete a high number of deals in a short time, I would travel down the learning curve faster than my peers, and this would allow me to become a strong candidate for a senior technology leadership position.

During the investment period at Intel, I co-founded the Intel optics division but decided early on that its chances of success were lower than I was comfortable with (it closed two years later). Soon after its launch, I departed the program. Having entered into the intrapreneurship that I walked away from at Motorola, I knew not to make the same mistakes and moved on to other things at Intel. These included mergers and acquisitions and focusing on treasury activities. Part of the corporate investor's job is to review business plans, which allows you to become astute at some of the inherent industry and technological problems. One of the major lessons learned while investing in start-up companies as an Intel employee is that whatever the business plan indicates, whether it is revenue, profitability, cash flow, technology, or something else, you invest in the team—the human capital is the most important. Business plans can and typically do change during the initial start-up period, and indeed change up to three times on average before a successful IPO. The team, if it is a good team, is flexible with its strategy, direction, tactics, and skilled personnel. Investment does not always go to the hungry entrepreneur; it goes to the experienced person or team that can show beyond a doubt that it will turn the investment into positive cash return. The best team has a better chance at securing funding for a technological project than a team with the best technology solutions but poor experience, no track record, and no organization. The combination of an experienced team, strong technological plans, and a decent amount of hunger generally has a great chance of success in raising capital, whether from banks, corporations, or venture capital.

The bubble was bursting, the investments were becoming more difficult, and the chance to do a lot of deals was evaporating quickly. I looked at my strategy again in 2000 and decided to test entrepreneurship.

5.6 Entrepreneurship

One of my fiercest competitors with OPTOBUS in the mid-1990s was employed at HP. With his team, we recruited him into AMP in 1999. When I left AMP to go to Intel, my colleague continued product development with AMP/Tyco and released industry-leading optoelectronics components. Tyco attempted to micromanage my colleague, who is a true intrapreneur, and he eventually left the company. He then came to me at Intel asking for funding to start a company.

After a number of discussions, it was clear that the skill base between us would be sufficient to pull together a world-class team and design a number of optoelectronics components that would technologically supersede anything that had been designed before. Our goal was to design and produce single-mode fiber optic datacom modules at the same cost basis as multimode modules. Historically, the ratio was 10:1; we had the team, plan, and expertise to bring that ratio down to 1:1. The venture capitalists liked the concept, and Ignis Optics (my first real entrepreneurial project) was born at the beginning of 2001. The first prototypes were released within 12 months, and first revenues were achieved within 24 months.

Unfortunately, the high-tech bubble had burst. The industry suffered from a huge surplus of parts, and the market collapsed. Even with the best sales and marketing folks at Ignis Optics, who had brought in marquee customers such as Cisco, it was clear that the company did not have the traction to wait until the industry achieved a reasonable equilibrium. Ignis Optics had industry-leading 10Gbps XFP fiber optic modules that were, in 2003, more than six months ahead of the major players and/or competition. The concern was that if Ignis Optics did not get significant traction toward positive cash flow within six to nine months, its perceived leadership status would wane and the marquee customers would simply go with suppliers with deeper fiscal resources. As entrepreneurs, we looked at our margins and concluded that we could give up a fraction to become a supplier on a sole-source arrangement. However, working with one major company on a sole-source arrangement is dangerous because you become dependent on its sales force. Putting in place four sole-source, nonexclusive contracts allowed the major players to compete against each other with the Ignis Optics product. We chose four companies who we felt had special relationships with their customers and whose customers tended to order in large quantities.

The concept of generating interest in Ignis Optics as a potential acquisition worked. We had expected that one of the four companies would envy the other three and want to acquire the company for itself. This indeed occurred, but unfortunately not at the value the founders and venture capitalists were hoping for. However, the return was 50 cents on the dollar, which exceeded all other optical acquisitions in 2003.

The entrepreneur days at Ignis Optics had finished, but the hunger was still there, and my overall strategy was still on track.

5.7 Summary

A common theme throughout my career is doing things many times over. This is how children learn: read the book or watch the movie again and again and again. Technologically, I did this by aggressively patenting new ideas, concepts, designs, etc., and again through the high number of investment deals at Intel. Both of these efforts allow you to approach the normal boundary conditions that are typically accepted in the community. Sometimes you can push the boundary a little further; sometimes not. In the end, the hunger to achieve a goal is what really makes the difference. Think logically, think on your feet, plan appropriately, attract key talent, become a trusted leader, maintain integrity, and above all, believe in yourself. It is the package of experience and integrity in academia, industry, finance, and business, with the hunger that enables the intrapreneur or entrepreneur to have a chance to succeed.

At the age of 45, I have accomplished a number of my goals in the optoelectronics business. The effort to stay on track with my original strategy is still alive; I've worked hard to keep to my original goals. I can honestly say that I still enjoy doing what I set out to do more than 30 years ago. That in itself is quite amazing. Most people change in their outlook and goals as they age, but it would seem that mine have not.

Subsequently, my tactics in supporting my strategy have evolved. I would not have anticipated a number of tactical disciplines had I sat down 30 years ago to plan this career in detail. I never expected to be involved in finance, corporate investments, or even innovative technology patenting, but I took the training just in case. Nevertheless, the experiences synthesize toward a goal which is as yet only partly achieved.

My next endeavor is to run an industry-based nonprofit trade association. The Optoelectronics Industry Development Association (OIDA) is allowing me to explore government relationships and tackle huge industry issues, concerns, and problems. The exciting thing is that these problems may be common and affect even the fiercest of competitors. Encouraging government to assist industry in certain areas is exciting, as well. Driving consensus in optoelectronics is rewarding; integrity, technological knowledge, and business acumen are critical for success. I feel honored that my career to date has enabled me to address global trade issues in optoelectronics and to network with the colleagues who can make a difference.

Part II

Some Case Studies

The aim of this section is to present the stories of 15 companies scattered through nine countries. All are involved in bringing research-based ideas to market, though through very different environments and involving very different personalities from a huge variety of backgrounds and interests.

All the stories in this section have much in common. All emphasize the critical roles of people within small teams. Trust, commitment, and flexibility are to the fore. Underlying this is also the unwritten faith that the founders of any operation must have in their instinctive judgment of people, circumstances, and structures. Business schools rarely figure. All the principals went through their first exposure to the process before the age of 50 and often under 40, though some of the addicts keep going to a ripe old age.

All are also initially based on a technology that the principals found and became experts in somewhere else, whether in university or in industry. This technological niche, this specialized knowledge, gives the initial edge in often highly competitive markets. All have also found the need to continue technology-based research to maintain their edge. Most, too, started with a single product idea and rapidly recognized the need to diversify, often into something very different.

The technologist also drastically underestimates the effort required to persuade the customer to write the check. This comes as a great surprise—after all, it is invariably self-evident to the technologists that their toys are the best on the block. To produce something with an assembly cost of $1,000 and sell it for $1,500 is a recipe for disaster. In these businesses, factors of three or more are necessary for survival and continued investment, and to cover marketing and numerous other overheads. Most of the products also need the international market. They are too specialized to exist within the local environment, and this international marketing, too, adds into the cost.

There are two other basic necessities. All recognize that entrepreneurial high-technology companies rarely, if ever, exist alone. They rely on partnership and inputs from elsewhere. Developing and nurturing these partnerships takes time and commitment, not to mention trust and instinct. Few of us can afford litigation fees when things go wrong.

The other essential ingredient is time. While there are occasionally instant success stories, going from rags to riches in a couple of years, more often the process involves a decade of dogged endeavor. This combination of risk, uncertainty, technical achievement, learning business acumen, making mistakes, establishing the network of contacts, and all that goes with it keeps the adrenalin flowing. All involved have, we believe, found the entire process immensely stimulating, very rewarding (sometimes even financially), and most of all, extremely enjoyable.

So these are the common themes. What about the differences?

Our 15 examples have their roots in very different sectors. About half are academic. Some are newly graduated students who have chosen enterprise. Others have evolved from industrial connections, though none, as far as we know, originated in government laboratories. This may be a quirk of the small sample, or perhaps it points to a trend.

Some consider the exit right at the beginning. These, in our sample, are a minority despite the common perception, and even frequent requirement, that a specified exit is essential for venture-capital funding. Indeed, you will see the funding models vary hugely.

Most of the companies have been founded on a product idea, though some have involved establishing agency or consultancy arms to support the finances. The products are diverse in character, many needing long acceptance, approval, and standardization rituals prior to becoming a source of continuing income. Most have also needed, to some extent, to create a market rather than simply address a current need, though very early in the process all have needed the ability to articulate the market to management boards and potential investors.

Our stories originate principally in the U.S. and Western Europe. We have only one (Fiberonics from New Delhi, India) from the developing world. China and India are viewed by most of us as potentially major technological contributors. The entrepreneurial environment has yet to develop, especially in the context of immense foreign investment, but develop it will. There is the instinct and energy at the individual level, already well recognized through entrepreneurial activity in the Western world, and this inevitably continues to mature along with the necessary supporting infrastructures.

We would not by any means claim that our selection spans all the options. Here you will see companies that gradually developed, examples of those absorbed by major corporations or holding companies with greater or lesser success, with the complex interplay of miscellaneous idiosyncratic funding models usually involving all of individuals, government and similar institutions, loan funds, and equity investment. You will see the common themes of commitment and enthusiasm and the immense variety of vehicles through which this commitment and enthusiasm emerge.

6

Mirada Solutions: The Case Study of a University Spin-Off

Miguel Mulet Parada and Sir Michael Brady
Members of the founding team of Mirada Solutions Ltd.
Professors, University of Oxford
United Kingdom

This chapter is a case study of a University of Oxford spin-off founded in 1999 and acquired in 2003 by a leading provider of medical imaging equipment. With this example, we hope to illustrate some recurrent themes of starting up a technology business in Europe and to offer a series of lessons learned that could be useful for future entrepreneurs.

6.1 The Origin

Mirada's origin goes back to two companies spun out in 1999 from the Medical Vision Laboratory (MVL) of the University of Oxford's Robotics Research Laboratory.

The MVL, founded and led by Prof. Mike Brady, started in 1991 with one doctoral student and grew during that decade to become a world reference in the field of computer vision applied to the interpretation of medical images. Prof. Brady already had a long career in robotics and computer vision, including the founding of an earlier start-up. In 1991, prompted by the death of a close relative from breast cancer, he started to work on diagnostic images. From this experience, he realized that a lot more could be done to help doctors make the most of diagnostic imaging by leveraging the use of computer-vision technologies.

In the 10 years that followed, the MVL extended its initial focus on mammography to cover the automated analysis of all major imaging modalities. The growing sophistication of imaging systems and the shift to digital data generated a great deal of academic research and prompted industrial interest and curiosity.

However, the clinical uptake of the new technologies seemed slow in the eyes of some MVL researchers who saw commercial opportunities for some of their tools. Spurred by the example of other successful Oxford spin-outs, two post-docs decided to set up separate companies to develop and sell their technology and asked Prof. Brady to become their chairman.

OMIA Ltd. was founded in late 1999 to concentrate on oncology and cardiology applications using magnetic resonance (MR), computed tomography (CT), positron emission tomography (PET), and ultrasound images. OXIVA Ltd., founded in early 2000, aimed to develop breast-cancer diagnosis tools using X-ray mammograms.

By early 2000, the two companies had obtained seed capital from the university and local business angels and had opened offices in a local incubator within a few minutes' cycling time from the MVL and the university. A year later, after completing a first round of funding, the two companies merged to become Mirada Solutions.

6.2 The Market Opportunity

The medical imaging market in 2000 was estimated at $30 billion in sales annually, 48% of which were in the U.S. The fastest-growing sector in this market is image-analysis software, which is also the highest value-added component. In a market in which competitors' products match each other's performance at breathtaking speed, software can provide crucial product differentiation. Better workflow design, improved diagnostic accuracy, and, in particular, access to novel clinical uses developed in conjunction with key opinion leaders can create a strong market pull for the products of the most innovative companies.

The larger scanner firms (such as Philips, GE, Siemens, and Toshiba) manufacture and integrate scanning systems that are sold directly to hospitals. These companies have a strong focus on hardware and its associated image-generation software. During the '90s, the progressive shift to digital imaging and the increasing resolution of volumetric data pushed scanner makers and traditional film companies (such as Kodak, Fuji, Agfa) to extend their expertise to image processing. They acquired or developed specialized technologies to render, manage, and interpret the digital data generated by their images.

In response to this technology gap, a first wave of entrepreneurs developed advanced applications for three-dimensional (3D) visualization and large-scale image storage and archiving systems. Some of these companies thrived on nonproprietary solutions, while larger players acquired the rest. By the time Mirada arrived, the 3D visualization and image-management niches were already pretty crowded and a second wave of innovation, offering smart image-interpretation software to help doctors in their diagnostic tasks, was starting to develop.

Mirada aimed to exploit this opportunity by filling the new gap between the generation of image data at one extreme and its visualization and storage at the other. The market gap was attractive because of three drivers:

- New diagnostic images display at a very high quality and detail and give rise to huge 3D datasets for which visual analysis may not be enough.
- Digital imaging and faster scanners reduce the cost of the examination and enable screening protocols that generate vast amounts of images that need fast diagnosis.
- The availability of radiological and nuclear-medicine imaging modalities and, in particular, the growth of PET imaging created the need to blend or fuse information from complementary examinations to obtain a full picture of the disease.

To fill this gap, Mirada positioned itself with three technologies originating in the MVL research:

- Fusion of multimodality images to blend CT, PET, and MR to support oncology diagnosis.
- Cardiac-wall tracking for echocardiographic exams (ultrasound imaging of the heart).
- Tumor detection in mammograms.

6.3 Image Fusion

Mirada's oncology project concentrated on fusing multimodality data. Typically, magnetic resonance (MR) and computed tomography (CT) provide anatomical images, showing structural differences that may indicate disease. Nuclear modalities [like positron emission tomography (PET) or single photon emission computed tomography (SPECT)] track the emission from a radiotracer as it accumulates in tissues, providing complementary information on the activity of diseased cells or tissues. When a patient is imaged using both modalities, the comparison of functional and anatomical data is a challenge because the patient lies in different positions in each scanner, the spatial resolution of the datasets may differ, and functional images display few anatomical cues to help the clinician align the two datasets.

Our initial versions of the software used rigid registration. This aligns volumes, applying translations and rotations after normalizing to the same resolution. Academics considered this a solved problem. However, academic papers mostly dealt with brain data, which is relatively simple to align, and described controlled conditions. Scanner manufacturers quickly highlighted that problems such as fusing a lung CT with a whole-body PET pose a massively more difficult generalization.

Moreover, the software needs to run continuously on many diverse platforms (e.g., imagine a PET machine on a truck backing up to a rural hospital and connecting over a serial line to a CT machine of any brand). The ability to run well-known algorithms on a sufficiently robust system proved an important technical challenge, ignored by the academics. Solving this opened a market niche, which proved crucial for the future of the company. This example illustrates the change of mindset that is necessary to grasp the intricacies of technology transfer and the tough, humbling work that is required to bring academic science to the real world.

6.4 Cardiac Tracking

Mirada's cardiology technology used ultrasound images to measure cardiac health. The assessment of wall motion using echocardiography remains a visual task, fully relying on the clinician's expertise and subjective judgment. Our technology fitted a "virtual elastic band" to the echo boundaries of the heart and analyzed its deformation to determine whether a wall segment was contracting poorly because of insufficient blood supply.

6.5 Breast Imaging

Mirada's breast-imaging technology hinged on a physics-based model that decomposed the intensity of the pixels on a mammogram into actual physical parameters corresponding to the amount of glandular or fatty tissue traversed by the X-ray. This representation, called the Standard Mammogram Form (SMFTM), enabled improved detection of breast cancer in screening protocols and more precise diagnosis. A software algorithm working on the SMFTM image could reliably extract features of interest that pointed to the presence of cancer, helping to make screening protocols more sensitive or determine whether the composition of the breast suggested a higher chance of developing the disease.

6.6 The First Year: Spinning out, Prototyping, and Seeking Funding

In a way, spinning out was the easiest step for both OMIA and OXIVA. Thanks to the support of the university and local business angels, the two companies were spun out relatively quickly. Teething troubles were more challenging. During 2000, the two embryonic companies faced many tests before they were finally born to the market as Mirada. OMIA and OXIVA focused internally on the development of commercially feasible prototypes while externally they sought potential customers and investment. As a way to compensate their burn rate and negotiate a first-round investment in the best possible terms, both companies applied for public R&D funding from all available sources. Despite the bureaucratic burden and some rigidity in its management, the public research money allowed for valuable breathing space and proved crucial for surviving this first year.

In terms of personnel, Mike Brady remained at the university while acting as chairman to the two companies. The original post-docs joined their companies in

CEO (OXIVA) and CTO roles (OMIA); while at the request of the university, a professional manager was brought into OMIA as CEO. Two recent doctoral graduates and two professional programmers (a pair in each) completed the core teams. In the case of OMIA, a doctoral student on a consulting contract and an experienced post-doc were lured from the U.S. and became Chief Scientists. These teams remained without changes during the course of the first year.

As increasingly professional prototypes were generated at OMIA and OXIVA, and development deals were struck with scanner makers, it became surprisingly more difficult to find a venture capitalist able to understand the technology and the market. Investors seemed split between those obsessed with the Internet and those beginning to drift away from technology companies altogether. Fortunately, a small VC fund seeking to specialize in biomedical technologies understood our two companies and decided to invest on the condition that OMIA and OXIVA merge. This merger gave rise to Mirada Solutions Ltd.

6.7 The Second Year: Finding the Route to Market

The investment allowed Mirada to move to a larger office in the incubator, buy proper equipment, and hire professional software engineers to create a professional development team as well as a vice president of sales and marketing (VPS&M) to lead commercial activities.

As a result of the merger, several roles had to be re-thought. Most important was that of CEO. Managing a small team of highly talented PhDs working on complex, uncertain projects on limited cash and few established processes is hard on its own. If you have to do this while seeking external clinical collaborations, convincing reticent customers, and seducing venture capitalists, it becomes an almost impossible task, very different from any other management function seen in the corporate world.

In our case, addressing these challenges required a great deal of teamwork and improvisation, and the understanding that the top roles and responsibilities needed rotating to tackle a new situation in which the external projection of Mirada was more important than keeping the house in order. The rigorous manager who is best at setting up the practicalities of the business may not be the best at dealing with the surprises of technology development, keeping VCs comfortable with daring strategies, or selling the idea to customers and investors across the pond. At this point, we learned how hard it is even for shrewd technology transfer professionals and investors to find the right kind of management talent for a start-up. A daring decision was made to appoint the OXIVA CEO, a post-doc with no industrial experience but who could grasp the science, translate the technology to market terms and lead the technical team, as the new CEO of Mirada.

With the cash in the bank to keep us going for another year, the focus during 2001 centered on taking our technologies to the market in a form that clinicians could use and would be willing to pay for. We soon realized that providing bespoke customized modules to big companies was not going to get us very far. These contracts were useful for validating our ideas, but development charges were modest, negotiations slow, payments took time to arrive, and customization laborious. Indirectly, the biggest benefit from supplying bespoke software was the brand recognition that it afforded Mirada when we started to sell direct to customers.

With the stand-alone idea in mind, the development team started to package the software with the right user interfaces, making it robust, obtaining clinical endorsement, and achieving the regulatory approvals that would enable the company to commercialize its technology in the U.S. and Europe, either directly or through scanner makers in an "Intel-inside" approach.

In addition to the technical work on the system, formal internal processes had to be introduced to substantially improve our academic approach to software engineering. Although regulatory requirements may seem a hindrance for small companies, the early implementation of a quality system helped Mirada deal successfully with its fast growth in the subsequent months. By the end of 2001, about 30 developers worked on three distinct product lines for various customers in a stable and controlled manner. This could not have happened without the proper quality and software development systems, which were put in place to meet regulatory and customer requirements. Interestingly, the cost of developing such systems was not prohibitive because Mirada counted on in-house talent to develop practical solutions, without resorting to expensive consultants or turnkey packages, and an industry-wise VP of Operations who, having been an MVL PhD himself, understood how to implement the process and lead the team forward.

In parallel with the development effort, the scientific team was reinforced with additional post-docs brought from the UK and abroad as well as various collaborations with scientific and clinical researchers. Among the new hires, Mirada found an amazing source of talent among graduate students applying for short industry projects. Surprisingly, these students always hired through trusted academic sources, displayed a deep knowledge of cutting-edge technologies, and proved adaptable to the hectic environment of the company. Again, an ability to learn on the job proved far more important than expensive expertise that was hard to validate or transfer.

As scientific work continued, a key area that demanded attention was intellectual property (IP) management. Isis, the university technology transfer office, provided much-needed support with technical advice and contacts with patent attorneys who helped execute our IP strategy. More important, the scientists previously dedicated to disseminating their research and absorbing others' papers quickly became adept at writing up patents and ensuring we were safe from patent infringement from competitors, proving yet again the need to adapt to radically different mindsets in the context of a start-up.

The same way that our VP of Operations and development team were key in building industrial-grade software from our patchy academic code, the immense, deep dedication and open mind of our chief scientist and his team enabled Mirada not only to develop novel algorithms worthy of many unpublished scientific articles, but also code that was robust and designed according to the best engineering practices. Thanks to the openness and maturity of all involved, there was never an R&D versus production divide. This seamless collaboration was so taken for granted that perhaps we never realized until now its contribution to Mirada's success.

While the internal focus was clear, Mirada had to face some key strategic questions in early 2001. In a start-up, strategy really boils down to agreeing on who your customers are and how to get to them as fast as possible; there is little time for more.

During this year, we spent many hours pondering questions like: Should our strategy focus on supplying branded "Intel-inside" software components to OEMs? Should we market stand-alone workstations to radiology departments or shrink-wrapped software to clinicians? Should we make alliances with re-sellers or go direct? Should we keep all our development lines open (mammography, cardiology, and oncology) or concentrate on one area? Should we focus geographically in either the U.S. or Europe or span both continents?

Of these questions, only one was easy to answer. It was obvious that to really make it, one had to succeed first in the U.S., where there was a more innovative customer base with access to cash and a very large and homogeneous market. Our VPS&M convinced everyone that this was the top priority and provided the leads to explore the market. However, the questions regarding our route to market and the product scope were harder to agree on.

The kind of strategic analysis that is valid in a corporate setting is simply not relevant in a start-up that is creating a new market and that has access to only incomplete information about its competitors and potential customers. Moreover, before thinking of who will buy its products, the managers of a venture-capital-backed start-up need clarity on who will buy the business down the line.

Gearing a company for an IPO may require a radically different strategy than preparing it for a trade sale or just plodding along in a niche (an option that is just not possible with venture-capital backers). The key question is not about tweaking a cash-flow profile to get a pretty curve over a number of years, but understanding why someone should buy the business altogether, who it might be, and then figuring out how to land the contracts that will prop up the cash-flow curve or the milestones that will increase the company's attractiveness. This involves having a vision for the company that all stakeholders agree on and from which an appropriate commercial strategy can be defined.

In our situation, as in many other cases, agreeing on a common vision proved difficult as a result of different incentives and perceptions among investors, founders, and management. Asking for a vision of Mirada in three years yielded radically different answers.

Fortunately, Mirada was lucky enough to remain flexible and attentive, allowing the market to answer this question for its management.

The most dramatic example of Mirada's strategic flexibility came during our first real foray in the commercial world of trade exhibitions. Following the advice of our VPS&M, it was decided that Mirada should have a presence at the Radiological Society of North America (RSNA) Scientific Meeting in November 2001. The RSNA meeting is Mecca for world radiologists. It was decided that Mirada would present itself as a serious player with staying power, and this required a respectable booth. The booth had to be big, eye catching, and located as close as possible to the big players. It had to be continuously staffed and show the best of our technologies. Our VPS&M promised to bring everyone who was anyone to meet Mirada if we managed this.

One can imagine how the idea of renting a booth that was about the size of our Oxford office resonated with a bunch of scientists who had never sold anything in their lives. RSNA was the riskiest decision in the life of Mirada, but also the one that provided the biggest return. In a week, the company spent one-fifth of the £1 million of capital raised. All the management and the development team plus some family and friends were flown to Chicago to staff the booth. Our VPS&M managed to land it right across from General Electric's booth, which at the time was our prime target and a pole attracting everyone's attention. The debutante was not going to be missed; the stakes could not be higher.

The booth contained five workstations showing the full range of Mirada's products with a core emphasis on mammo, which we thought at the time to be the most attractive and innovative piece of technology. Demonstrators were available around the clock. With a bit of training and a lot of enthusiasm, everyone became a salesman in record time, and any passerby who showed any interest was pulled in for a software demo. Demonstrators were busy all the time, while the management team managed to meet with corporate and medical opinion leaders nonstop to explore potential deals.

As the exhibition hall closed on the first day, the team met for a postmortem and voiced their impressions. We had all been busy, but not as expected. A theme started to emerge; customers were not really dying for our mammo products just yet. The image-fusion package seemed the big thing in the show and interest correlated with the new hybrid PET/CT scanners on display. Many clinicians and more than a few engineers from the competition had come to have a look. Mammo was attracting many researchers but not that many potential customers. We had two days left, and perhaps we needed to change our tactics.

Based on this feedback, we quickly decided to turn the booth around, change the video loop at the welcome desk, and dedicate four of the five workstations primarily to the fusion package. On the second and third days, the booth became even busier and the level and depth of the commercial discussions increased. Companies came around asking to arrange feasibility tests of the tool and clinicians wanted to have it on their desks as early as possible.

It seemed some of the questions had been answered. The team returned to Oxford with a lot of homework and a clear picture that fusion was going to be the next big thing, and Mirada could be part of it.

6.8 The Third Year: Building the Business

Mirada came back from Chicago with a firm order to supply a large scanner maker with a complete fusion solution, including hardware that it would then resell directly to its American customers. We had to quickly strike deals with workstation vendors and develop a self-contained, fully validated system with full regulatory approval to ship in early 2002. This deal led to our first real sales, partly compensating the big investment in the show. In addition, various key opinion leaders at top U.S. hospitals requested their own research workstations, and serious commercial conversations started with other large scanner and 3D visualization vendors who also needed to fill their gap in fusion.

During this time, it was decided to discontinue the cardiology project, despite various successful feasibility projects with the major ultrasound manufacturers. The acquisition of these companies by larger players put a stop to their interest in external technologies, and the development effort required to go it alone seemed excessive in the light of the fusion priorities.

The more difficult decision involved the mammography project. This had been a pet project of the MVL, the origin of OXIVA, and an area in which we believed we were unique. However, the pressure to deliver a fusion solution meant that resources needed reallocation. Although there was some demand for computer-aided diagnostic software, we were aware that significant efforts were still required to achieve full clinical validation and educate customers on the use of a novel diagnostic approach. On the other hand, users of PET and CT were already asking for something that could accurately fuse the data from their scanners. They had a very clear picture of what they needed and how to use it. This decision was particularly hard for our CEO, who had founded OXIVA and had dedicated more than 10 years of work on mammography; yet he realized the importance of maximizing the company's short-term chances and fully backed the decision to put mammography on ice while shifting focus to fusion.

The mammography work was not completely stopped. Thanks to two large, publicly funded, collaborative research projects, development continued while licensing negotiations started with a manufacturer of computer-aided diagnosis systems.

Having clarified our focus, the company concentrated during 2002 on getting the new Fusion7D workstations out of the door. A presence was maintained in the major shows, and further deals with end customers and corporate companies were agreed on.

Just when the outlook seemed brighter, Mirada ran into difficulties. Cash management is key in a start-up, where there is very little slack to deal with late payments and practically no negotiation power to pursue favorable payment conditions from suppliers. It only took one of our customers to delay its payment to place Mirada in an extremely difficult situation.

During a few hectic days, the founders scrambled to persuade our investors and others to inject some further cash into the company. This involved difficult negotiations and frictions, but the final agreement bought us a few more months to build our revenues and think of a serious exit. Once this glitch was over, it was time to re-address the vision based on our newly found market knowledge.

It was obvious that there was a gap in image-understanding expertise among scanner manufacturers, and that the interest in filling this gap stemmed from the strong growth of their PET businesses. Combined PET/CT scanners producing the two datasets in one examination had become successful. Yet to fully exploit these machines, it was necessary to align the data very accurately, a task that required a software solution. In addition, the prospect of PET/CT imaging went beyond the existing tracer (fluorodeoxyglucose) used to illuminate fast-metabolizing cells. The expectation that new, tumor-specific radio-tracers could be designed revealed a huge potential market. A new field of molecular imaging was being created not only to revolutionize diagnosis but also to open new avenues for joint diagnostic/therapeutic systems for clinical use and pharmaceutical research. Scanner manufacturers saw an opportunity to build a new business around molecular imaging, including radio-tracer design, specific imaging protocols, and quantification software.

The new vision for Mirada came as a corollary to this insight. Mirada would become a molecular-imaging company devising the technologies that enabled the design and use of software-enhanced pharmaceuticals and in this way become a prime acquisition target for the large scanner manufacturers who needed a foothold in the new field.

Our business-development VP quickly grasped the idea and aggressively championed it inside and out. As a former MVL PhD, Chris Behrenbruch had sound technical knowledge and a deep commercial insight nurtured by his close collaboration with our VPS&M. Chris' strong drive to sell the idea outside Mirada earned him the role of CEO towards the end of 2002.

Chris quickly followed up with our potential customers at RSNA 2002 and got an appointment to discuss our new workstation design with CTI, the company that invented PET.

6.9 The Exit

Events quickly unfolded after our second RSNA visit. Following the conference, Chris and Mike went to meet CTI in the U.S. in January 2003. After a presentation of Mirada's capabilities, track record, and its fusion project, a grueling but inconclusive question session followed.

Back in Oxford, just as they were starting to wonder what would happen, Mike and Chris received a call from CTI's senior management. It turned out that CTI was not interested in our Fusion workstations.

The disappointment was short lived. CTI was actually interested in the company and was ready to make an offer for it. Our vision and strategy had somehow worked out! The CTI and Mirada boards started negotiating the terms of a sale immediately and after a few weeks, a final proposal for discussion with Mirada's shareholders was agreed upon in March 2003.

Receiving a firm offer for the company is no guarantee of a sale unless all the stockholders reach an agreement. Different incentives and outlooks mean that disagreements on the "right" price for the company may crop up between founders and investors. In the case of Mirada, the period of internal negotiations was especially tough for those who had a personal stake in the business. Seeing uncertainty looming behind what seemed such an excellent solution was frustrating. What could be better than obtaining a tidy sum on top of merging into the recognized leader in the field, with whom we already had excellent collaborations?

Resolving this negotiation took a long time, but a final agreement among all parties was finally reached six months later. According to this agreement, the company was sold to CTI for $23 million. CTI would keep the Mirada team in Oxford to become its software arm while our CEO became managing director of the new CTI Mirada. Integration into CTI followed relatively smoothly.

As Mike put it in 2004, Mirada became the sinews of the new CTI organization as the supplier of its imaging-software solutions. Besides the internal transfer pricing agreements with CTI, Mirada maintained commercial contracts with other companies, most notably the mammography CAD company R2, Vital Images, a 3D visualization company, and radio-tracer maker Amersham.

The deal was not only positive for Mirada's founders and investors, but the University of Oxford, which invested £150,000 of seed capital obtained £1,070,000 plus £400,000 in royalties. Part of this money helped to build a new premise for the MVL at the university, where Mike Brady continues to be professor while considering future ventures such as Ixico and Dexela.

6.10 Lessons Learned: Conclusions

The aim of this chapter was not only to give an account of a real case in starting up a company but also to provide some sort of pointers that could be helpful to future entrepreneurs. To do this, we have avoided an academic discussion of what is just an anecdote of successful entrepreneurship. Trying to be practical and humble, we have aimed to condense our recollections into three lists in which we hope readers can find some relevant tips for their own circumstances. The first list includes those things that in retrospect worked well, the second includes some issues that could have killed the company, and a final list enumerates some lessons that we think we will take with us in future ventures. Many lessons will probably seem hard or even impossible to implement, but then you will also need a large dose of luck (and effort) to bypass these shortcomings.

6.11 The Things That Worked Well

During start-up in Oxford, the following were invaluable:

- Isis Innovation: The university's technology transfer company provided invaluable help in the first phases, including access to funding, networks, and advice regarding IP management.
- The Oxford Innovation: The local incubator provided affordable office premises and shared services within cycling distance of the university, allowing us to remain close to the MVL and keep commuting cheap and short for the first few cash-strapped employees.
- The Oxford entrepreneurial environment in general: Oxford provided frequent contacts with other entrepreneurs, the evidence of other success stories, and access to professionals accustomed to working for or with small companies, including bankers, lawyers, and many of our first few employees outside the lab.
- The friends, fools, families, and business angels who bet on the first two companies and on Mirada when it was close to bankruptcy.
- The Oxford name: Sometimes it does count where you come from. Oxford helped us open some doors among investors, customers, and collaborators. However, note that once the door is open, everyone is equal.

Management decisions that proved beneficial included the following:

- The tough cash discipline imposed during the first year by our first CEO to make the seed funds last until viable prototypes were ready.
- The good use of public research funding that provided a cash cushion on which to develop our products.
- Hiring a street-wise VPS&M with huge experience and an incredible agenda.
- Appointing the right people at the right time and changing CEOs as the company needed them.
- As a corollary to the above point, having a team that was able to cope with the changes and recognized their strengths and limitations.
- Having a stable chairman throughout in the person of Mike Brady to ensure continuity. Even if we were to agree with him that his technical input was negligible, his role in keeping such diverse, big personalities on the same path or at least not bringing the whole edifice down was far more important than the gray-haired scientific savvy advertised in the business plan that was so highly prized by investors and customers!
- Understanding that the future was in the U.S. and betting the house on RSNA 2001.
- Understanding that we were there to sell our technology, not to explain how fantastically clever it was and how badly doctors needed it.
- Knowing how to tell a story well… and knowing how to tell a different story if necessary.

- Keeping all options open, including going directly to the final user with research workstations that gave us visibility and access to real market intelligence and know-how from key opinion leaders.
- Being flexible in terms of strategy and roles and responsibilities as well as having a team that, despite its differences, was able to work together and keep the fantasy/reality factor under control.
- Finding our niche and finding CTI.

6.12 Things That Could Have Killed the Company

- The moody world of VCs before and after September 11, 2001, when New York City suffered a terrorist attack.
- Time wasted strategizing on a business plan beyond what was plausible and useful.
- Some unavoidable personality clashes throughout the life of the company—be ready for them.
- Cash-flow management and the lack of negotiation power to manage accounts payable and receivable.
- The long love affairs with various OEMs that played hard to get while keeping us locked in feasibility projects.
- Expecting our OEM customers to tell us what the business was about and not asking the end-users early enough.
- The lack of a common vision with regard to an exit strategy among our shareholders and founders during our growth phase and misalignment of incentives during the sale negotiation.
- A very complicated shareholding structure resulting from the merger of OMIA and OXIVA and issues with the early distribution of shares that was not commensurate with the effort brought by different parties to the company.
- The temptation to try everything, promise anything, and deliver nothing.

6.13 Some Recommendations

- Build a sound story about the company rather than the most precise business plan. Focus on what makes it unique and why someone would buy the company as well as its products.
- Understand the structure of a term sheet and get the advice of a good corporate lawyer.[1,2]
- Understand what will trigger the strategy of your VC in different situations and be able to anticipate it.
- Choose a VC who brings more than money. Make sure they understand start-ups and your sector and are willing to work hard for you. Ask them

what they are offering. Remember, it is a two-way exchange, so ensure that there are mechanisms to keep them to their promises.

- Be very careful with the initial allocation of shares. They are very expensive rewards. Make sure they are matched to long-term commitment to the company. Ensure that share-option schemes provide the right incentives and are useful to bring enough of the right people into the company because you will not be able to compete on salaries.
- Hire people within trusted networks. You cannot afford to make mistakes. If you cannot hire the perfect candidate, it may be better not to hire. If the candidate fits the team and has a strong ambition to learn on the job, even if experience is just below what is required, hire that person. Fast learners are better than know-it-alls. Most often, you cannot judge past experience, but you can tell a clever person willing to take on challenges. In any case, don't hire too late or fire too late.
- If you hire from the corporate world, make sure you understand why these people left that environment. This is much more important than what they supposedly did in a much cozier environment than that afforded by a start-up. Ensure you know why they would like to work for you and have a share in the future (whatever it might be) of your company.
- Wisely manage IP and use it as a negotiating tool. Use patent searches as a cheap source of market intelligence.
- Focus on cash flow! Strike the best possible payment terms. Don't waste money on experts, reports, or superfluous things. Borrow and beg where possible. Cash in the bank is your best negotiation asset.
- As soon as you can, hire a good financial director with experience in start-ups or VCs.
- Hire a brilliant VP of Business Development with sales experience and the fattest phone book.
- Realize that in many circumstances, the American market (or even going global) are not options, they are imperatives.
- Even if you deal in a business-to-business market, it pays to go directly to the final user to learn from them. The OEM sale is not necessarily the shortest route to market.
- Try to enlist in your team a visionary, a brilliant scientist, a rigorous technician, an aggressive salesperson, a cautious manager, and a natural leader who can make all of them work together.
- Find the CEO who most closely matches all of the above. Consider whether changes or rotations are needed. Keep stability with a respected and active chair.
- Remain realistic, sell what people need and want. There is no time to convince them of the next best thing. You are running against the clock and your current account.

- Talk to all stakeholders all the time, including shareholders, founders, investors, employees, and customers.
- One never aims too high; you only have one chance at a time (but remember you can always set up another start-up in the future).

References

1. Pearce, Rupert and Barnes, Simon, *Raising Venture Capital*, John Wiley & Sons, Ltd. (2006).
2. Campbell, Katharine, *Smarter Ventures: A Survivor's Guide to Venture Capital through the New Cycle*, Financial Times Prentice Hall (2003).

7

Building a Company the Old-Fashioned Way: Meadowlark Optics, Inc.

Tom Baur and Garry Gorsuch
Cofounders of Meadowlark Optics, Inc.
Frederick, Colorado, United States

This chapter presents the path followed by these entrepreneurs to build the new company. Considerations concerning the business plan, the financing, legal considerations, and funding principles at the beginning, followed with comments about the formative and transitional years, are included. The chapter concludes with key topics to be considered by entrepreneurs planning to start a new business.

7.1 The Beginning

When our company began in 1979, it was a one-person, part-time effort in a spare room in my home. There was a single product, a wide field-of-view Pockels cell that served a tiny market of astronomers making polarization measurements. It had no marketing or sales effort, no rent to pay, and a total capital investment in manufacturing and test equipment of less than $1,000. This was a time of learning the fundamentals of a manufacturing business, and the order stream was certainly slow enough to allow time for me to learn to walk before I ran. I named the enterprise Meadowlark Optics because it was located on Meadowlark Hill, and because it began as a lark in a meadow on my ranch.

Doing business was fascinating, and it was a thrill for me, a researcher in a federally funded lab, to put my skills and funds to work in a way in which the link between performance and reward was much more direct than in a large research laboratory. Filling a marketplace need, however small, was exciting. I coasted along in this hobby mode for about four years, bootstrapping my way to more and better equipment, but still with only one product and one small group of customers.

Everything changed in 1983, four years after we began. Changes at my "real" job left me with a single remaining employment possibility involving a move from Colorado to Maryland. The move to and challenges in Maryland were less appealing, and I succumbed to the siren song of entrepreneurship, where I had found the risk-reward equation to suit my temperament.

7.2 The Business Plan

What to do? Mortgage and car payments could not be extracted from this tiny enterprise. Growing the company was a must for my personal financial survival. This was exciting! There was much to learn and critical choices to make. We needed more products. We needed a real marketing and sales effort. We needed time to develop these products and equipment to build them efficiently. In short, we needed ca$$$h! And we needed a business plan that defined our direction and goals. From this plan came the cash and personnel requirements to execute this plan.

The plan centered on building a technology-based company that specialized in precision optical components for polarization control because that was the technology that I knew best. The plan opened doors for financing growth. Writing the business plan forced me to learn the rudiments of marketing and sales, cash flow, and the importance of applying the technological know-how I possessed to building products the marketplace needed. Building “neat stuff” that no one would buy would lead to disaster. The plan called for products to serve laboratory needs because this was really the only market I knew at all. I had no grandiose plans to sell large quantities of product to any market. In other words, I chose a niche strategy.

I had no thoughts about exit strategy and only foggy notions about the need for profitability beyond what was needed to fund cash-flow needs and some modest growth.

7.3 Financing

Because I chose a niche strategy, our early funding needs were modest. I had a bootstrap financial model for growth through profit leveraged with debt. Initial equity financing was less than $30,000. I borrowed about $65,000 through a State of Colorado loan guarantee program similar to the Small Business Administration 7(a) program, literally betting the farm on the success of this venture. Also, I borrowed about $60,000 from friends. This modest financial “war chest” was adequate because we sold only component-level products into small niches and because photonics technology advanced more slowly in the early 1980s. There were fewer large markets to drive rapid growth. The bootstrap mentality fit well with my only other business experience, which was in agriculture. It would not fit well for most high-tech start-ups today because these start-ups generally plan to bring their technology to market in a large, fast-moving market, often with fierce, well-funded competitors.

7.4 Legal Considerations

The company was incorporated as a C corporation in 1979 but switched to an S corporation in 1981. Today, there are other viable choices as well. They include limited liability corporations and limited liability limited partnerships. These choices depend on the investors' tax situation and on the exit strategy for the investors and founders. If sophisticated investors provide equity financing for a company, they will dictate the choice of organization entity. These entities can provide some degree of insulation for the stockholders from liability for any actions by the company. Business insurance is needed for protection of the company's assets, not only from physical threats such as theft and fire but also to protect those assets against product-liability claims, lawsuits from disgruntled employees, breach of contract suits, patent infringement suits, etc. Every business must navigate through a minefield of government regulations and legal perils that can quickly bankrupt any well-meaning entrepreneur who has not built an adequate defense of correct legal structure, adequate knowledge of government regulations, good legal counsel, and good insurance. One fine for violating an export regulation could mean the end of your company. It is not enough to have honorable intent. You and your advisors must know the laws and understand the legal exposures resulting from all company decisions and policies.

From the beginning, I had a vision of the kind of company I wanted to build. The excitement of building a business organization from scratch was a bigger motivator than any dreams of riches in the distant future. I wanted the business to be something we could point to with pride. It had to align with our personal values and be built on integrity of product, integrity in relations with customers, in dealing with financial backers, and especially integrity in relationships with employees. Our products must meet specifications; we must be honest with our customers and our financial backers, and we must keep our promises to them. We had to build a company that is a good place to work, where all employees are valued and treated with respect.

We wanted to be known for quality products and quality customer service. Products must always meet promised specifications, and customers must have the technical assistance they need to make the best product selections.

The company had to be financially sound. It had to be a good investment for our backers and strong enough to provide continuity of employment for those who joined me in this venture. We needed enough profit to grow and continue to develop new products to assure that the company had a bright future.

7.5 The Formative Years

The early years after Meadowlark Optics moved from a hobby business into a full-time occupation in 1983 were the most exciting for me as the founder. The process of building a going concern from nothing satisfied my creative urges. The risk and uncertainty excited me after 13 years in the relative safety and predictability of a large research institution.

The first few months I was a one-man band, keeping the books, developing new products, making products for sale, shipping, receiving, developing sales literature, and developing advertisements. The list was long and so were the hours. Soon, I learned that just doing business took more time than making a product, so I hired an office manager and three manufacturing technicians. Our "burn rate" went way up and put real urgency on getting new products to market before our cash was gone.

A year later, we blew by the breakeven point and began to turn our attention to building a complete company team. We hired an engineer/salesman and more manufacturing help. We retained a full-service accounting firm to help with financial matters. We moved the business from my house to the renovated barn on my ranch. Business grew, and soon we remodeled a large chicken coop to give us more space. Even the company pet pig was nervous about losing his pen to this enterprise.

These early years were not always positive. Our sales went up and down with the economy. We had some very painful contractions in the early 90s when the economy was down. The survival of the company was in doubt, our bankers were nervous, I lost sleep, and some very good, dedicated employees lost their jobs.

Still, the overall trend was upward. By 1994, we had more than 30 people managed by an astrophysicist with no previous experience in business other than farming and ranching. We had little organizational structure, poor procedure documentation, loose or no job descriptions, inadequate delegation, and no clear lines of authority. This was a wreck waiting to happen, and the entrepreneur (me) was getting way out of his skill set in managing the company. I was becoming the bottleneck restricting the company's growth and was not temperamentally suited to putting in place the structure and organization we needed for further growth. We brought in an experienced manager to help us make the transition to this new level in 1995. This move was critical to our successes since then. I believe that recognizing your own weaknesses and hiring to shore them up is vital to entrepreneurial success. If your enterprise is backed by venture capital, these investors will help force this recognition on you.

7.6 The Transition Years

The time at which a company passes from a start-up to a more mature organization is not easily defined. It comes down to the complexity of handling many business activities at the same time, and one or two people cannot stay on top of all of them. When decisions can no longer be made in a timely manner and information needed to make those decisions is not readily available, it is time to look at some changes. An infrastructure to handle the business and technical processes that run a company is in order.

For us, the magic number from a personnel standpoint was around 25 people. Although the normal personnel practices have to be in place for any number of people, there is a point when tracking and reporting complexity gets too intensive to effectively manage.

The number of orders, shipments, part numbers, and financial complexity drives a decision to install a logistics system. This can be very painful for any business, and it was for us as well. It required not only installation and support of a new system but training in use of the system. The discipline required to keep the databases in sync with what was happening in the business was a major effort. People who were used to picking out a part in inventory now needed to fill out the paperwork needed for costing, ordering, and inventory control.

Pricing was turning into a complex problem. In the past, the head engineer or the CEO could decide on the price. The question of what products make money as well as what products were losing became critical. We found that our intuitive sense was not always correct. We had to turn to our logistics system. We chose a system that was mid-range in capability with minimum complexity. The process of implementing such a program is never ending. The cost of both support and operation is high, but the resulting control is a necessary requirement of running a business.

Skills are a continuous area of focus. The size and demands on the business were such that new skills and talent needed to be added. An introduction to some candidates through mutual friends resulted in the hiring of a general manager and a VP of sales and marketing.

As the business grew, priority setting became a more important consideration. Where should we place our efforts? Too many things were happening to make decisions based on gut feelings. It now required some criteria to decide what, when, and whether things needed to be done.

Documentation also became more important. The ability to run the business verbally was degrading rapidly and some form of communication vehicle was needed. A more sophisticated part-number system was developed and put into the logistics system. Procedures in support of formal processes needed to be developed. An International Organization for Standardization (ISO)-type formal process was started, but the main focus was on processes and documentation in any form.

The company was operating in about 4,000 square feet, and the need for better facilities forced us to look for a new facility. We started by looking at space to lease and could not come up with a satisfactory solution. The decision was made to buy land and build. The loans to buy and build were arranged through the U.S. Small Business Administration's 504 program.

It was a huge advantage to start with a blank piece of paper and design a building that would meet our needs now and for the foreseeable future. We decided on a steel building with a pitched roof on a three-acre parcel of the purchased land. We designed clean rooms, a chemical room, assembly, engineering, and administrative space. The land was laid out such that the Meadowlark owner could develop a future business park. Meadowlark leased the building, which ended up at 20,000 square feet, with about half being manufacturing clean rooms.

One key element of the whole building process was the cooperation of a friendly local government that valued our company's move into its city and county. Decisions were fast and reasonably in agreement with what we needed. In the summer of 1997, we moved the company over one weekend. Through detailed planning and execution the business was up and running within a week; in a month, most processes were up to speed.

The location and building gave the business dramatic results, but the timing proved to be a problem. As we were adjusting to the new overhead structure of our building, the semiconductor industry took a hit, and some of our largest customers were forced to cut back. This resulted in some trying times for the business. However, the move was essential, and future work depended on our facility, both from a size and a utilitarian standpoint. Meadowlark Optics emerged from the downturn stronger than ever.

Some personnel challenges arose as the company became more structured. Some people could not adjust and left the business. Others were skeptical of the changes. New people were hired to meet the needs of the business as some of the former employees left. In order to develop a spirit of teamwork, an incentive plan was put in place. If the revenue goal for the month was met, everyone in the company got a paid day off. There was no question about making the goal; it was how to make the goal.

The company had always recruited and hired highly talented people, and in the engineering ranks we found ourselves hiring PhDs for most of our technical leadership. The hires were of high quality and resulted in refined processes, better response to customers' technical questions, better quotes, and product improvement.

Internally, we took great pride in our products and abilities, but we needed to find out what the outside world thought of us. A market survey verified that our company was known for its high product quality, technical capability, and most of all metrology. Our strategy was to continue to focus on these strengths.

Marketing and sales were a key effort. We developed a catalog. Our reputation carried the business a long way, but our advertising had been modest and limited. We needed to project a more technical image to our potential customers. We redesigned the company logo to give it a more modern look and strengthened ads to be more colorful and eye catching. We put much more effort into our Website as well as our visibility in the marketplace. We also strengthened our relationship with SPIE, the technical photonics magazines, and our customers. We emerged with a new look and strengthened our commitment to customer support and satisfaction.

Safety was always an important area, and the desire to improve resulted in naming a safety coordinator and initiating a formal process. Employee awareness was and is key to a good safety program. Colorado State University, which has an Occupational Safety and Health Administration (OSHA) support program, helped us with audits and suggestions on how to improve meeting OSHA guidelines. Chemical handling also was a focus item, and the new building helped us provide new, improved controls.

The company had always kept stringent quality controls in place, and our reputation for expertise in metrology was and still is a major strength. We needed to extend the same rigor to the business operation as well as continue to improve our quality processes. An advantage the business had was that everyone in the company had been immersed in the company philosophy: never knowingly shipping a bad product. The product had to meet the customer's expectations. They understood the importance of quality and our reputation. Compromise was out of the question.

Off that base, we began the initial steps toward process control. They key word was "measurement." A saying we used repeatedly was, "What gets measured gets done." Charts started showing up all over the company. Deliveries, sales, cleanliness, on-time deliveries... you name it and there was a measurement for it; we continue to this day. And, by measuring, action plans can be implemented, and the understanding of our business and processes becomes apparent and can then be improved.

We continually improved our financial process. Our customer base expanded to include more OEMs, increasing our volume and abilities. We had some government-related projects and audits that were handled effectively and with positive results.

7.7 The Growth Stage

Even though we have a catalog and sell some standard products, our main market is custom polarization optics. Our customers work with us to adapt or design new parts that meet their needs. Manufacturing volumes of the parts comes some time later.

The industries we support are telecommunications, medical, biotechnology, semiconductors, research institutions, military, and space applications. Like every company, we are at the mercy of the economy, and by spreading our industry base we hope to avoid deep industry cycles. Although this is an interesting goal, in practice we get caught up in growth industries more than in stable or declining industries. Our desire is to build up financial strength during the good parts of the economic cycle to carry us over the down cycles.

The key in any phase of a company is the skills it possesses in its employees. While in the start-up phases, there are usually several key people who possess the skills needed to keep it going; as a company grows, more and a larger variety of skills are required.

Developing these skills in present or finding new employees who have the requisite skills is a vital requirement and challenge. As we grew, we needed to expand our sales and marketing group in number and expertise; likewise in research and development. We needed new skills in order to expand our product line and respond to customers' requests for new products. Management development was a challenge because we needed to separate our tasks into departments and functions.

Our understanding of business cases and market requirements increased or at least became more sophisticated. The balance sheet and profit and loss (P&L) statement became a tool for more people to use as decisions were pushed down in the decision chain. Financial process improvements were continually upgraded, and the focus on cash management, accounts receivable, and inventory as well as overall financial control continued to improve and the company grew.

Engineering document controls evolved from an informal, relatively uncontrolled process to a formal process requiring a full-time document-control person with special skills in level control, tracking, and distribution. On the quality front, which had traditionally been the major strength of our company, we discovered that although we put out a high-quality product, we had to improve our quality process. We again turned to the ISO guides for direction.

Engineering had to learn project management involving several individuals and projects. A small company does not have the luxury of single assignments, and individuals must handle a wide variety of tasks. This learning experience is an ongoing one and will probably never be complete.

A continuous effort is directed toward better planning. Everything from cash management to product development to training technicians needs to be beefed up. This will continue as a requirement for running a successful business.

Any company that is subject to the effects of economic trends must try to foresee the events that will affect it. It also must take action to survive as a business if the economic situation affects it negatively. In the 2000/2001 technical economic downturn, the company had to take some undesirable actions. Personnel, salaries, and expenses were reduced. We all hope that this is the end of actions that will need to be taken, but the future is murky in the high-technology arena.

As the economy improves and demand for products increases, the company will emerge stronger. New products under development will expand our customer base. Customers are designing optical components that need our kind of product with the support of our technical expertise. The business plan of the past will continue to function in the future, but the company must and will be ready to change and alter its direction if need be.

7.8 Some Considerations When Starting a New Business

Considerations that were critical in the past to starting a business are changing, and the new business environment is going to be more challenging for the small business start-up as well the established small business. Although many of the following apply to any company, these comments are more relevant to a bootstrap start-up business venture.

For a new engineer, starting a business may sound glamorous and have an attractive element of freedom. And that can be. However, make up your mind that starting a business is mostly finding funding, managing finances, and managing your financiers. We chose to use a few loans and not rely on venture funding. In today's world, that is almost unthinkable.

Remember that the only reason a company exists is its customers. Focus on their wants and needs, respect their objectives and desires, and understand their businesses. Networking has and always will be a key to your success.

In the long term, your company must make money. That means that you need not only a unique technology but also to be focused in managing people and the business. A couple of years ago, making a profit was considered kind of a quirky idea. That mindset disappeared in 2000 and 2001 as companies with no profit outlook were in trouble.

There are two basic management tools: the P&L statement and cash planning and management. For a new company, cash management will always be the most important tool. As the time comes for moving on to an IPO, merging or somehow making your investors a return, the P&L statement must look good.

In the end, the only thing your company will have is the people who work and support it. They will add to your intellectual property, your products, and your processes. You will have to multiply your skills, desires, and plans through people. The better they are and the more skills they have, the better the business will be. This sounds so simple, but you will find that you will become more dependent on your employees; soon, there is no business without them. Hire carefully and treat them well so they will stay. Loyalty is a scarce commodity in today's world. You will have to earn it.

Good personnel practices, wages, and benefits are essential and, while they can grow with the business to hold good people, you must find a way to meet your employees' needs both financially and professionally.

Ever heard of OSHA, the EPA, Worker's Compensation, employer liability insurance, or the NLRB? These and many more acronyms and organizations will enter your life as soon as you hire someone. Keeping these and other administrative details at bay while providing a stable work environment will eat up more of your time than you can ever imagine, but they are essential. While you might find someone to handle the paperwork, the decisions are yours alone.

Developing the market for your product is critical, for if there is no sale, there is no company. A sale is the most important step you can make because without one your future is very limited. If you are a technical person without sales experience, hire someone who has the experience and listen to them. It takes a unique technical person to understand marketing and selling.

A piece of advice I was once given holds pretty well for most companies: make money on everything you do. That means don't bet "on the come," making big investments on the promise of big things to come. Large volumes, grandiose stories of how well a potential customer is funded, etc., can provide a huge incentive to develop a product for them, only to see them drift away after you have invested a large amount of time and money on the potential opportunity. Make sure you understand their market and keep a cool head when analyzing the opportunity.

Know when, and do not be afraid, to say no. When cash seems to be at the door, don't get overly wrapped up in emotion, or soon you and your resources can be buried in an effort that may not be profitable. That's called digging yourself a bigger hole. Survival drives people to do things that can make matters worse.

Keep a distance between your personal and business attachments. Doing a favor for your buddy can many times lead to overcommitting and putting your people in a compromising position with other customers. When the backyard barbecue discussion turns to "Can you do this?", pull out the order sheet and fill in all the blanks. Explain that this will fit in with everything else, for a price. Give them a smile and get on with flipping the burger. The result of accepting an order that turns out to be a losing proposition can hurt team morale as much as the financial loss you might suffer.

Remember the earlier saying, "What gets measured gets done." Measure everything, analyze every part of your business, and have a constant improvement plan in place.

Develop a set of principles and stick to them. This sounds easy but takes a great deal of thought and dedication. Explain them to your employees, your customers, and your vendors. The Golden Rule may sound old-fashioned, but it usually works.

Before you bite off the challenge of starting a new business, take some finance classes and talk with people who have done it—not just the successful ones, but also some who have failed. They will gladly give their advice.

After considering all these things, take a big breath. If you think you're ready, go for it. You can always go back to work for someone. But you'll probably never be happy working for anyone again.

8

Building a Lasting Optical Design and Manufacturing Company

Jay Kumler
Coastal Optical Systems, Inc.
Jupiter, Florida, United States

Aviator Charles Lindbergh said, "Success is not measured by what a man accomplishes, but by the opposition he has encountered, and the courage with which he has maintained the struggle against overwhelming odds." To that definition, we would add, "Success is also measured by the fun that you have with great people in the process."

What factors have enabled Coastal Optical Systems to compete successfully? We try to find the right people, and we concentrate on those things that we do best. The success our company has achieved can be credited to the dedicated, hard-working team at Coastal. Each day we try to be a little better at what we do. This chapter will describe some of the lessons we have learned along the way.

8.1 Background

Coastal Optical Systems grew out of the disintegration of the U.S. defense industry in the early 1990's. Observers have written that difficult economies and times of market change are the best time to start a new company. The great downsizing of defense-related photonics from 1989 to 1991 was the period of market upheaval that birthed Coastal Optical Systems.

Our business plan (written in 1991) outlined our strategy for growth:

- Evolve from 100% aerospace and defense to 75% commercial and 25% aerospace.
- Transition from 100% optical components to 100% precision lens assemblies.
- Help customers rapidly take new designs from the drawing board to hardware (optical design, mechanical design, fabrication, and test under one roof).

For the first 10 years, the company had the same leadership and ownership. Together, we managed to flourish in a four-way partnership. Observers were sure that there could never be so many bulls in one china shop, but we turned a disadvantage into an advantage. Our business decision-making process has been enhanced because every operational meeting is also a de facto board of directors meeting. Having team ownership allowed us to make decisions faster, not slower. It has allowed us to use each other as sounding boards with free exchanges because all four owners set aside individual interests and worked in the best interests of the company. Coastal is an example wherein four heads are better than one.

From 1994 to 1996, Coastal struggled and only grew at a compounded annual growth rate of 11%. We were working hard, but we were not very profitable, and sales were not growing. As a team, we had to reevaluate why we went into business and what we were going to try to accomplish with our efforts.

8.2 Five Keys to Coastal's Early Success

From 1997 to 2002, Coastal grew at a compounded annual growth rate (CAGR) of 22.6%. Just as important, our EBITDA (earnings before interest, taxes, depreciation, and amortization) doubled. No one defining action marked the turning point for Coastal Optical Systems. We achieved improved performance and increasing momentum by concentrating on five strategic points: focusing on what we do best, broadening our customer base, teaming with like-minded customers, hiring people with our company's core values, and treating our employees like family.

8.2.1 Focus on What We Do Best

Of the 700,000 new businesses started each year, only 35,000 (or one in 20) will be around five years from now. The primary reason for a small company's failure is trying to do too many different things at once.[1] For the first five years (1992-1997), Coastal was in this situation.

As a start-up manufacturing company, the tendency was to take any work to keep machines running and pay the bills. During the first few years, we would work on anything. We polished stainless steel inserts for progressive ophthalmic lenses, shaped prisms, polished germanium hemispheres, core-drilled cover glass, cut plastic lenses, and diced cylinders. Within a couple of years, we had lots of activity and lots of long hours, with little profits and slow growth. We had done little to build lasting value into our company. We were forced to start focusing our business. We had to decide what we were passionate about, what we could be the best in the world at, and what brought us the most profits.

Coastal Optical Systems has always been a manufacturing company. We invent and build hardware. We are not a research firm. We are not optical design consultants. We do not pursue small-business innovative research (SBIR) opportunities. Our product is hardware on the table, not paper. Coastal Optical Systems is passionate about designing and building world-class lens assemblies. We named the company Coastal Optical Systems 15 years ago because we envisioned a company that would excel in designing and building lens assemblies. Coastal Optical Systems was our name and also our objective.

Marketing strategist and writer Al Ries writes, "A focused company puts its best people and most of its resources into the products or services that represent the future. While it may have to deal with products from the past, the focused company makes no bones about the fact that these are not in its future plans."[2] We hope that a simple focus will establish us in the mind of the customer: Coastal Optical Systems equals world-class lenses. This focus on lenses also helps us develop a shared mindset in our company.[3] We want our employees to think "world-class lenses" almost as much as we want our customers to. Everyone on the Coastal team needs to understand that we are passionate about custom world-class lenses.

8.2.2 Broaden Our Customer Base

In 1994, 67% of our gross revenues came from a single aerospace customer. As suddenly as a South Florida storm, the customer decided it would no longer purchase lenses from us. Losing two-thirds of our business in our third year of operation was a shock that could have sunk our company. In retrospect, it was the best thing that could have happened. It forced us to broaden our markets and look for customers. It forced us out of the complacency that arises when one customer is providing you with large amounts of work.

Since 2000, the largest single customer accounted for only 13% of our total sales in any one fiscal year. It is a learned fundamental that our company must sell its products in as many markets as possible. Coastal sells lenses to four markets:

	% of sales in 2007
Aerospace and defense	*24.7%*
Digital imaging and projection	*28.4%*
Automated optical inspection and metrology	*25.5%*
Biomedical applications	*21.4%*
Total	***100%***

The traditional precision-optics industry in the United States has always served cyclical markets. Aerospace, microlithography, industrial manufacturing, and telecommunications are all highly cyclical industries characterized by unpredictable cycles of boom and bust. Coastal sells to four global markets—when one market is down, the other markets pick up the slack. The year 2001 was a down year for microlithography and machine vision and saw the evaporation of the telecommunications marketplace, but Coastal grew 20% in 2001 on the strength of aerospace and digital projection.

Coastal's design teams can work on a consumer 35-mm photography project one day and a satellite-based hyperspectral imager the next. Selling lenses to multiple markets makes good business sense and makes Coastal a more interesting place to work.

8.2.3 We Align Ourselves with Like-minded Customers

We have sought opportunities to team with customers whenever possible. Teaming with the right customers during the design phase allows us to draw from all disciplines, including optical and mechanical design, fabrication assembly, and testing. For a small company, strategic teaming agreements also protect our intellectual property and protect us from being sold out from an ongoing product because the customer bid out the volume production to a lower bidder.

Over time, as we concentrate on quality and stay focused on our core product, our company gravitates towards commercial customers that have technical requirements consistent with our focus. These commercial customers understand the mutual benefit of design and development teams. Combining our customers' expertise in systems design, product design, and market needs—with Coastal's understanding of optical engineering and fabrication—brings better products to market faster and more efficiently. Vertically aligned teams of designers and fabricators are more responsive, more efficient, and can design and build products faster.

In aerospace, the traditional bid-and-proposal cycle has nearly disappeared on custom projects and one-of-a-kind satellite instruments. Our aerospace customers no longer have the luxury of receiving competitive bids from five or six different qualified suppliers. This is because the precision optics industry has lost many of the companies that traditionally supplied space-qualified optical systems (Itek, Perkin Elmer, and United Technologies Optical Systems, for instance.) This is also because of the ever-increasing demand for lower cost and quicker development of space instruments and missions. Successful aerospace companies have decided to align themselves with technology partners that team from proposal to design to fabrication and integration. Aerospace prime contractors who recognize this market change realize that they can no longer be competitive with arm's-length customer/supplier relationships.

We have been fortunate enough to team with two or three key companies in each of our markets. Coastal has had the opportunity to work on a remarkable range of technically rewarding projects because our customers share our attitude towards open and honest teamwork.

8.2.4 We Hire People with the Same Core Values

According to best-selling author James Collins (*Built to Last* and *Good to Great*), core values "are the organization's essential and enduring tenets: a small set of timeless guiding principles that require no external justification; they have intrinsic value and importance to those inside the organization."[4] At Coastal

Optical Systems, our core values are honesty, hard work, and humility. These core values are so near and dear to us that we use these as a guide for our business and personal lives:

Honesty: All of our relationships will be handled with honesty. All business with suppliers and customers is conducted with the highest level of integrity. When we are honest with our customers, they are honest with us.
Hard work: After 15 years, we still get up early every day and work hard. People who do not see the value in honest, hard work are not happy working at Coastal. Designing and building world-class optics is hard work.
Humility: We have tried to create a workplace where unselfishness is valued and where teamwork is the name of the game. Each employee will do whatever it takes to get the job done. We don't care who gets the credit.

In his book *Clicks and Mortar*, David Pottruck writes that, within limits, he tries to "hire for intelligence, character, and attitude, and then train for skill."[5] Coastal Optical Systems has always hired for attitude and trained for skill. The right people don't need to be tightly managed or fired up; they will be self-motivated to produce the best results and to be part of creating something great. Finding people who share our core values and with whom we enjoy working side by side takes time and effort. There is no litmus test and no easy screening process. Instead of the traditional one-hour interview, we try to spend extended periods of time (two days) with prospective employees in our shop so they experience our work environment. Then, they spend a weekend in the area so they can decide whether South Florida is a place where they want to relocate their family. After this extended period of courtship, if we are still in doubt, we keep looking.

8.2.5 We Are Generous with our Employees

Coastal Optical Systems understands that the limit on growth for many companies is not markets, or technology, or competition, or products. It is the ability to get and keep enough of the right people.

We try to create a culture that rewards disciplined people and recognizes our core values. For example, our lunchroom has a Wall of Fame displaying engraved plaques for each employee who has been a team member for five years or more. The personalized plaques describe how each key employee personifies Coastal's core values. Each year, we look forward to adding another employee to the Wall of Fame with a special presentation at the company Christmas party in front of the employee's family and coworkers.

We trust our employees. We try to create a company around the idea of freedom and personal responsibility within the framework of our company. The employee will go to great lengths to fulfill his responsibilities. In return, Coastal rewards personal responsibility by going the extra mile. Whether it is paid time off so that they can see a sick father, or extended grace when an employee is having difficulties at home, Coastal treats disciplined employees like family. In the end, the mutual trust pays off many times over.

Coastal Optical Systems has also tried to supply the best in benefits. Generous health insurance, 401K retirement plans, profit sharing, bonuses, and vacations make good business sense and create a close-knit team.

8.3 Challenges We Face as a Company

The photonics industry in North America continues to evolve and react to shifts in the marketplace.

The greatest challenge the photonics industry faces is competition from China and other countries with lower manufacturing costs. North American companies must determine their value proposition and how they will compete in the world markets. Coastal has been sourcing select components overseas since 2002, and we will continue to work with overseas partners in the future. The greatest risk is to our intellectual property. Even when we take every possible precaution, our camera lenses have been reverse engineered and reintroduced to the North American market by Chinese manufacturers who have no hesitation stealing proprietary designs. We are prepared to compete on the world market, but the current playing field is not level.

The second challenge is the weak dollar. Coastal imports infrared thermography cameras from Germany and sells the cameras in the United States. European products are now 40% more expensive in the United States due to the changes in the exchange rate over the past three years.

The third challenge is the restrictions imposed by the International Trade in Arms Regulations (ITAR). Since 2001, ITAR restrictions have greatly restricted U.S. companies from exporting a wide range of products, including high-power lasers, diamond turning machines, infrared cameras, and night-vision equipment. The U.S. restrictions appear to penalize U.S. companies unfairly, and the regulations rarely keep up with the constantly changing technologies. ITAR limits the subcontractors we can do business with, adds legal costs, administrative costs, and makes doing business more and more challenging for small companies.

All of these challenges are also opportunities. A well-planned strategy for buying and selling in China can grow our company. A weak dollar means that Coastal's products are more affordable in Europe. ITAR restrictions limit the number of companies that can manufacture restricted hardware to U.S. corporations in some cases.

8.4 Changes in Photonics Offer New, Exciting Opportunities

There has never been a better time to be involved in optics. Photonics technology has captured the imagination of the world and has received hundreds of millions of dollars of investment capital. Some (like digital photography and digital projection) started 10 years ago but continue to strongly influence Coastal's products. Other emerging markets are only recently leaving the research laboratories and universities but show promise for the future.

8.4.1 Digital Photography

Since the invention of film-based photography, there may never have been a period of technological advancement in photography like the digital photographic revolution. According to the 2001 PMA U.S. Consumer Photo Buying Report, 1.7% of still cameras purchased by respondents at camera stores and one-hour photo labs were digital in 1997. That number increased to 18.3% in 2000. According to InfoTrends Research Group, worldwide consumer digital camera sales will hit 82 million units in 2008. Coastal is concentrating on the "prosumer" and professional digital camera market by supplying high performance specialty lenses for special digital photography applications like immersive photography.

8.4.2 DLP-based Digital Projection Technology and Electronic Cinema

The Digital Micromirror Device (DMD) is a reflective array of fast, digital light switches that are monolithically integrated onto a silicon address chip. Digital Light Processing (DLP) projection display systems based on the DMD provide high-quality, seamless, all-digital images that have exceptional stability and freedom from image lag. The DMD switch has an array of up to 2.2 million hinged, microscopic mirrors that operate as optical switches to create a high-resolution, full-color image. Texas Instruments supplies DLP subsystems to more than 75 of the world's top projector manufacturers, who then design, manufacture, and market projectors based on DLP technology. More than 10 million DLP subsystems have been shipped since early 1996.

Theaters have installed digital projection systems on more than 5,000 screens worldwide. Coastal Optical Systems is designing and building custom projection lenses for digital theaters and has also used DMDs to build custom projection devices. New projection technology means new requirements for increasingly higher performance lenses.

8.4.3 Multispectral and hyperspectral imaging

Scientists and researchers have long known that a tremendous amount of important information is outside the visible spectrum. As sensors for long-wave infrared, mid-wave infrared, near-infrared and ultraviolet become more and more cost effective, the demand for broadband and multiband color corrected optics increases each year. Multispectral imaging is used in agricultural research, border security, remote sensing, astronomy, forensic sciences, and biomedical applications. All of these applications need custom, high-performance imaging optics for spectral bands outside the visible spectrum.

Multisensor image fusion is the process of combining information from two or more images into a single image. In medicine alone, multispectral image fusion is used in MRI, computed tomography, position emission tomography (PET), and radiology. Coastal is working on providing infrared thermography images for many of these applications.

8.4.4 Space Science Enterprises

NASA continues to chart the evolution of the universe from origins to destiny and improve our understanding of galaxies and planets. Even with the constriction of the NASA budgets, NASA is pursuing an expanded robotic program to explore the solar system and universe. The Lunar Reconnaissance Orbiter is scheduled to launch in the fall of 2008 to map the surface of the Moon and search for future landing sites. The Mars Science Laboratory will launch in 2009.

The optics industry plays a central role in earth-observing missions, Mars exploration programs, and Discovery missions—the low-cost planetary missions with short development schedules. One example is Deep Impact, which fired a copper projectile into the comet P/Tempel 1 to expose the comet's interior ice and rock. This enabled the spacecraft to perform the first-ever study of unaltered cometary material. Coastal Optical Systems supplied satellite optical subassemblies and components on many instruments currently in orbit, including the Space Telescope Imaging Spectrograph (STIS) and the Near-Infrared Camera and Multi-Object Spectrograph (NICMOS). Coastal continues its tradition of supporting NASA missions as a current supplier to NASA working on lenses for the James Webb Space Telescope NIRCam instrument.

8.5 Conclusion

The rapid changes in the photonics industry offer exciting opportunities for a focused company. By concentrating on world-class lens assemblies, we have a strategic plan to team with key companies in many of these new markets to develop new products based on new enabling technologies. We are disciplined enough to concentrate on doing only what we do best.

Coastal Optical Systems' success is attributed to focusing on what we do best, broadening our customer base, teaming with like-minded customers, hiring people with our company's core values, and treating our employees like family. This has helped us profitably grow at a 27% annualized growth rate from 2003-2007.

Six years after merging Coastal Optical Systems with the global photonics leader Jenoptik Laser Optik Systeme, we still concentrate on the core purpose that we established years ago. This clear focus helps serve as our compass as the ownership of the company changed, founders of the company retired, and the optics market in North America evolved. Operating practices, customers, products, and markets change, but our core values are unchanged.

References

1. Ries, A., *Focus: The Future Of Your Company Depends On It*, Harper Collins (2005)
2. Ries, A., *Focus: The Future of your Company Depends on It*, Harper Collins, p. 280 (2005)
3. Ashkenas, R., *The Boundaryless Organization*, Jossey-Bass Management Series, p. 71 (2002)
4. Collins, J. C. and Porras, J. I., *Built to Last: Successful Habits of Visionary Companies*, Harper Collins, p.222 (1994)
5. Pottruck, D. S. and Pearce, T., *Clicks and Mortar: Passion-Driven Growth in an Internet-Driven World*, Jossey-Bass Business & Management Series (2001)

9

The Life and Times of a High-Tech Entrepreneur

Colleen Fitzpatrick
Founder and President of Rice Systems, Inc.
Irvine, California, United States

Twelve years ago, when I was laid off from my job at a small high-tech laser company, I decided to go it on my own. I figured that if nothing else, I would at least learn something. Little did I realize how right I was. Because women in laser physics are in the vast minority, and female high-tech business owners are an even smaller group, if nothing else, I have learned what it is to persevere.

9.1 The Beginning

In 1989, I started my company in my garage—or, more accurately, in half of my garage (I still needed a place to park my car). Although I grew up around businessmen—my father was a successful entrepreneur—I had little if any practical business experience. At that time, I started my company because that is what I felt was the next step. The company I had just left was disintegrating, which brought out the worst in everyone as they scrambled to protect their jobs. I did not want to walk back into a cut-throat atmosphere like that with the same kind of boss. I wanted to disconnect from the fast track that had so blindly swept me away and reorient my life to go more in the direction I wanted it to go.

When I lost my job, I received no severance package, so I had only my own savings to live on. I thought I could last about a year without an income, but actually, my savings ran out long before that, and I lived on loans until I received payment on my first contract. During the first couple of years in business, I did whatever I could to earn money to support bad habits such as paying my bills on time. I tutored high school students in math, I did word processing, and I even went on game shows. (To answer your next question, "Wheel of Fortune" and "To Tell the Truth," where I won a combined total of five house payments.) In my spare time, I decided to enjoy life, so I had inexpensive adventures such as white-water rafting and volunteering as an extra for the local opera. I also made my home as comfortable a place to work as possible, since I was spending the majority of my time there.

Many entrepreneurs will tell you that they started their companies with great ideas that they were going to sell, or a product or a customer they inherited from someone else. Not so in my case. I started my company out of the necessity of making a living. All I had was some basic knowledge of my profession and a small optical laboratory that I had built in my garage over the years. I knew a few people, but I can't say I was a recognized name in the field.

What I did have was a bit of resourcefulness, a natural ability to meet people, and very good communication skills. Actually, I owe my successful start to a colleague who visited my tiny garage lab one afternoon. I told him I was considering bidding for a small business contract. By coincidence, he had decided to bid the same one, and we discussed our various approaches. To be honest, I thought his was ridiculous and mine was at least that good. So I wrote the proposal, drawing together the skills of the ex-secretary, the ex-accounts manager, and the ex-contracts negotiator of my former company. I also obtained a proposal from an ex-program manager who had been at this company sometime in the past, a man who had won a high percentage of the contracts he bid.

I read that proposal over and over for six weeks and milked it for every tiny shred of psychology—how many equations to include, how complex the figures could be, the flow of logic, the level of technical explanation, and how all of these elements fit together into a whole. Since I did not have a computer of my own, I carried all my reference material to the media division of the local library and rented one, first-come, first-served, for $3/hour. I realized that writing and submitting proposals was an essential skill, whether I worked for my own company or for someone else's. I had nothing to lose and a lot to gain, whether I was awarded the contract or not.

Six months later, when I received the results in the mail, I opened the letter just to see how a rejection sounded. Well, if you heard a scream coming from Fountain Valley in 1990 or so, it came from me when I found out that I had been chosen for award on that contract. I was so excited that I could not even remember what I had proposed to do!

9.2 The First Contracts

This first contract gave me a chance to learn some valuable skills on my own, without the umbrella of more senior managers and scientists. I learned to talk to customers, to have the courage to call people I didn't know and talk to them about my technology, where to find ideas. It wasn't easy. I continued winning one or two small-business contracts per year, continuing to work in my garage. My friends used to kid me that defense contracting had become a cottage industry.

As the sophistication of my work grew, so did my need for lab space and equipment. Soon, I was renting a lab in my old company, which had since downsized and had extra space available. I also rented a lab at a nearby university with a pulsed laser. My big break came in late 1994 when I was awarded two NASA small-business contracts. This allowed me to hire my first employee, who could now do the lab work required by my contracts. This released me to do what I do best—writing proposals, meeting people, and coming up with new ideas. Finally, in mid-1996, six years after I had started out, I rented my own commercial space with a seven-figure company and eight employees.

Sounds easy, doesn't it? Well, let me tell you a few more details to illustrate just how hard things can be. The real story is that in mid-1994, I was going broke and running out of ideas. I realized that I did not quite understand the business side of things, and that, in spite of my trickle of contracts, I was losing ground. I had to make the decision whether to continue or not. After a period of intense soul-searching, I dug in my heels and borrowed what at that time was a large sum of money. I stayed up late, I wrote proposals, sometimes all night long. I worked hard. The award of those two NASA contracts was the result of this commitment combined with a bit of good luck. I realized that not only was I a hard worker, but I was also pretty lucky, and that the harder I worked, the luckier I got. One of those NASA contracts led to the award of an SBIR Phase II contract.

A couple of other events turned out to be critical. A colleague of mine lost his job at Hughes. At that time, he was consulting with a small company that was developing an integrated optical gyroscope. However, his new employer would not allow him to participate in the project, as they felt it might be a conflict of interest. He called me and asked me to substitute. Months later, as a result of my involvement in this program, when the original small business was no longer able to continue with the project, it was transferred to Rice Systems, leading to the award of a second SBIR Phase II. All together, as a result of my renewed commitment to my company, I had four contracts at the end of 1994 (including an additional consulting contract I had picked up in the meantime), leading to two Phase II SBIR contracts by the end of 1995. I was doing better than ever.

9.3 Still Sound Easy? Hang with Me a Moment

Because of these two SBIR Phase IIs, in a virtual sense, the company was worth seven figures. However, at that time the government shut down because of a budget crisis, so that the SBIR program was essentially frozen until further notice. In addition, some legalities were involved with the transfer (novation) of the gyroscope contract to Rice Systems. The net result was that I had bills to pay, including the salary of one employee, with a lot of funding in the wings but no immediate income and no savings to draw on. Practically speaking, the company was worth nothing.

The only thing I could do was to finance both my company and me from the second mortgage on my house and hope that I could last longer than Congress could argue. Finally, in the spring of 1996 the government resolved its crisis, the legalities of the novation were complete, and we were ready to negotiate our two Phase II contracts.

The catch was that before signing the contracts, the government asked me to show I could carry the company for 60 days, as I could not submit my first invoice for 30 days, net 30. At that time, I had exhausted all my liquid assets. I had about $1,000 left in my second mortgage, with a house payment of about $l,000 due within a couple of weeks. I applied to three banks for a short-term loan, but they would not take the two Phase II contracts as collateral. If I were financing a washing-machine store, fine, I could use my company's receivables as collateral but not if I was a government contractor. Therefore, I was in the position that I could not get the loan without using the contracts as collateral, and I could not get the contracts without the loan. In the end, my family cosigned a short-term loan with their bank to keep me for the first 60 days of the contract, which added to the debt of the first loan from 1994 and my (now depleted) second mortgage. But I definitely knew I could make it with the two new contracts. On March 1, 1996, my company went from nothing to a seven-figure company in one day. By the way, the financing provided by this new loan lasted almost exactly 60 days, without an ounce of fat. (See Rule of Thumb No. 3 below). It took me several years to recover financially.

As I stated above, in mid-1996, six years after I had started out, I rented my own commercial space with a seven-figure company and eight employees. Does it still sound like it was easy?

When I eventually looked back at the steady growth of my company over the years, in spite of all these setbacks, I began to have confidence in the future and believe that I was really doing something right. I began to notice that when a crisis appeared on the horizon, the resolution was somehow already in the works, and just when the answer was needed the most, it emerged from the past to catch up with the present and saved the day. I knew that while I could not count on this happening all the time, I could afford to have a little more confidence in the long-term survival of my company.

While I have had many colleagues who also formed their own companies during the years, I have seen most of them fail. On the surface, the reasons varied from arrogance to incompetence to just plain bad luck. However, I am convinced that a more fundamental reason lies behind these more obvious explanations. I believe the true reason for failure is the inability of these business owners to listen. When I consider the particular individuals in question, the common denominator is that they had a strong personal agenda that eclipsed their ability to understand what they needed to do to satisfy their customers.

In one case, a colleague was determined to separate his home life from the responsibilities of his small business. He did not want to fall into what he saw as the trap of bringing his work home with him. He wanted to leave work at a reasonable time in order to spend undistracted evenings at home with his wife. This is, in fact, in conflict with the quick turnaround time and the high return on the dollar that are two of the major advantages of dealing with a small business in the first place. I approached him about a subcontract to do the engineering on a laser system I was bidding. His company was a partnership involving one other individual and his own wife. This contract could have made his company.

I needed his help immediately as I worked night and day to prepare my proposal, but two weeks went by and I still had not heard from him. When I urgently telephoned him on the Friday before the Tuesday deadline, he informed me he was in process of preparing his estimate of the work, and that since his wife was already gone that day, he would discuss it with her first thing the following Monday morning. (Wife? Monday morning? Didn't he live with her and couldn't he call her at home and do it right at that moment? Didn't he see her every day and then in the evening, too? Why hadn't he already taken care of this in the last two weeks? He had a new company, and I knew that he was not busy.)

Eventually, when he did give me something the following Monday, having lost faith in his ability to deliver, I had already worked around what he could have supplied me and had gutted his write-up for the remaining missing items. He went out of business a short time later, losing all his retirement savings. I am sure he still has not figured out why.

To balance this example with a success story, the need to listen (and the relief that comes with being listened to) was again emphasized for me recently on a visit to a company in Oregon to market a type of light source we had developed. It was a busy day for the owner, yet he took the time to examine my light source carefully and to perform one or two simple experiments to show the capabilities and limitations of the source. In the middle of the busy morning, we finally settled down to talk about our respective companies. Although his company was not high-tech, he impressed me as an intelligent man who listened intently as I described some of the other projects we were working on in nondestructive testing, telecommunications, micro-electro-mechanical systems (MEMS) sensors, and security applications.

His cell phone rang, and he quickly answered a few questions for the caller. When he hung up, he picked up our conversation with the exact sentence I had spoken at the moment his phone rang. The conversation went on as if it had not been broken. I was impressed. He is definitely a person I want to do business with—because he listens.

The ability to write well is also critical but also largely ignored. This is especially frustrating because of the large number of individuals in our industry who do not speak English as their first language. In my company, for example, more than half my employees over the last five years have not been native-born Americans, representing every continent except Australia and Antarctica. I cannot express my frustration at their lack of writing skills. It is not so much the inability of an employee to write and communicate effectively that bothers me; it is mostly the inability of the person to recognize the problem and ask for help. I have always tried to help my employees improve their writing, but very few of them have understood how critical this is to their success.

There is a comment I'd like to make that might seem out of context here, but it is very important. It is something that my father always told me: It is a lot harder to be No. 1 than No. 2. One of the most difficult things I have had to face on a regular basis is that many of my employees, as well as many of my colleagues, believe they know how to run my company better than I do. But in reality, they don't have a grasp of the scope of responsibility involved with maintaining a small business. Some only see the freedom to control others and to call the shots, while others believe that owning a company is an automatic guarantee of a huge income without any strings attached.

The truth is that owning a small business is like being the captain of a small boat on a rough sea. Sometimes you ride the waves and sometimes you think you are about to sink, but whatever happens, you are the only one who can steer the boat. While I think I am reasonably aware of the opinions of those around me, and often incorporate them into the decisions I make, it is still for me to make the decisions. My employees can always get another job. I can't.

I have one ex-employee who is probably still angry that I fired him. He did not agree with the way I was running the company, and his behavior completely impeded any course of action I thought was necessary. Although he was actually destroying the company, he came to believe that only he could run it effectively. He thought he was the guiding force of the operation and that I was completely powerless to keep the company functioning without him. So I fired him. He made the rounds in the community with claims that he was solely responsible for the success of the company, and that now that he was gone, it would disintegrate without him. It is now several years later, and Rice Systems is still going strong. My last knowledge of this ex-employee is that he is either unemployed or else working for a big company where he is, in fact, almost powerless. He has professionally disappeared.

Being a woman in the high-tech world has both advantages and disadvantages. Along the way, I have come to realize that something that is both a key asset of and a liability for a high-tech businesswoman is what I call "unidirectional visibility." One day, when I walked into a meeting related to a large government program in which I was involved, I scanned the attendees to assess the tone of the talk I would give. I saw about 30 men in attendance. Suddenly, it occurred to me that from my point of view there was a relatively uniform population in the room. That is, I saw 30 men. But anyone else in the room would see something different—29 men and one woman. There was no way I could not be noticed.

Many times, I will call a program manager who is associated with a proposal I'd like to submit. Often, when I mention my name, he will respond, "Oh, I know you. Don't you remember I met you at the so-and-so meeting last January? We spoke in the hall for 20 minutes after the session was over."

Of course, I will have at most only a vague recollection of who he might be. So I respond, "Oh, yes, I remember you. You were the fellow with the dark suit and the red tie, right?" In most cases, this is enough for me to sneak by. This is only one example of how I profit from my unidirectional visibility.

Visibility can also be a disadvantage, the cause of great discomfort. Occasionally, I attend crowded meetings where I am one of only a few women. To overcome my nervousness about not only being a stranger in the crowd, but more important, being so noticeable, I single out one or two women and introduce myself. I usually say something like, "There aren't many women at this meeting, so I always like to introduce myself to them." At that point, I have an ally of some kind, and it is easier for me to feel comfortable with the rest of the crowd.

Over the years, I have developed three rules of thumb I use in my business, which could just as well be applied to life in general. The first is staying in business is surviving your mistakes. This implies that there is no way to avoid making them; it's just surviving them that is the important element. If I thought my last boss was difficult, the one I have now (myself) is even worse at pointing out everything I do wrong. Yet, allowing myself to recognize the mistakes in the first place and a chance to think my way through them gives me a chance to learn from them and go on.

Unfortunately, I have a massive library from which to draw for an example of this principle. I have made mistakes in negotiations that cost a great deal of money. I have made mistakes in hiring employees. I have made mistakes in the strategy used to position a critical proposal. Yet, I have managed to land on my feet (so far). When I was first starting out, I took my mistakes more personally in that I did not have the experience to do otherwise. But one thing I did that helped me carry on when I was particularly discouraged was that I made a notebook of articles that I clipped from newspapers and magazines highlighting women who had overcome overwhelming obstacles to become successful in their fields.

Did you know that Bette Davis made 22 movies before she won her first Academy Award for Best Actress for *Dangerous* in 1935, and 14 additional movies before she won her second award for *Jezebel* in 1938? She was nominated for the Best Actress Award 11 times, including a write-in nomination for playing Mildred in *Of Human Bondage* in 1934.

Another article that caught my eye was about a young woman from Los Angeles named Lupe Vasquez who had just won a full scholarship to Stanford. The interesting part was that this woman's family was homeless, yet she had not been discouraged by her circumstances and had worked as hard as she could for a good education and a scholarship.

Still, a third article that caught my attention was a short paragraph in the Milestones Section of *Time* magazine. It read, "Expecting. Benazir Bhutto, 36, Prime Minister of Pakistan, and her husband, businessman Asif Ali Zardari, 36: their second child; by year's end. If Bhutto survives a vote of no confidence planned by the opposition for this week, she will be the first 20th century head of government to give birth while in office."

I felt a kinship with these women who had struggled so hard and realized that I was not alone. Through these and other articles on accomplished women, I realized that success is not easy, and that the right attitude was to take everything that happened, good or bad, as a learning experience and move on. Since then, my secret ambition has been to compose an inspirational book about women like these, both famous and not, devoting a chapter to the struggle that each one experienced leading up to her ultimate success.

My second rule of thumb is if you don't feel peace about it, don't do it. Often, I am surrounded by circumstances that demand an immediate decision. However, if I am aware of an uneasiness as I decide what to do, I have learned to trust my intuition that, on some level, something is wrong, and I must stop a moment to consider the situation in more detail. Often, I cannot pinpoint the reason for my hesitation, but if I wait, it will become obvious.

I developed this principle early in my career while I was on the science faculty of a university in Texas. I was a dedicated teacher who worked very hard at developing coursework and labs for my students, yet at the end I realized that I was not being paid for my contributions to the department, and the administration was dragging its feet on promoting me to a promised tenure-track position. I interviewed for and was offered jobs at several universities in various parts of the country, but none that was any better than the position I held at that time. In the end, I turned down all but two of the most promising. I knew they were my last chance for a ticket out.

As the time approached to make a final decision, I felt trapped. If I accepted either job, I would escape from my unhappy situation only to start at the bottom of one that could be even worse. But if I turned down these jobs I regarded as unappealing, I would be stuck where I was. In the course of agonizing over what I was going to do, one of my relatives told me something that only registered hours later. He advised me, "If you don't feel peace about either job, don't accept it." I took his advice.

I was miserable for the next six weeks while I tried to justify to myself what I had done. What I had forgotten was that almost as a joke, I had given my résumé to one of my former students who had just landed a job as a member of the technical staff with Rockwell (now Boeing) in Seal Beach, Calif. I had never thought seriously about making the change from academia into industry. This student had been a housewife rearing two sons. When they were old enough to be in school most of the day, she had gone back to community college and picked up algebra at the "one plus one equals two" level and had worked her way into a cum laude bachelor's degree in physics at the school where I taught. When she arrived in California, she passed my résumé on to someone in management. The rest is history.

The third rule of thumb: to stay in business, you have all you need and you need all you have, and there is not one ounce extra in either direction. This is true in a number of ways. If I am working in the lab doing a complicated experiment late at night before a deadline, I will have just enough '/i-20 screws to tie down all the optical components, I will have just enough change in my pocket to buy that last cup of coffee, and I will finish five minutes before I have to get the report to Federal Express the next day. It never fails, no matter how early I start, no matter how much money I got from the ATM that afternoon, and no matter how late Federal Express closes. If I could write a novel instead of a short article, I would still not exhaust all the situations where this rule of thumb has held true, so I will limit myself to just two examples.

During my earliest days in business, I had to watch my finances. Not only was I starting my company, but I had to live. I was trying to keep the balance on my credit cards low and not fall into the trap of charging everything I needed, only to run into heavy debt when the bills came due. One month, my balance was

$350, a little higher than I felt comfortable with. I had no alternative other than to dip into my dwindling savings to pay the bill. When I went to collect the mail a little later, I received two letters. One was a refund from the IRS for about $332, from an error I had made three years earlier in rounding off my tax return to one too many decimal places. The second was a check from a friend who had stayed at my home over Christmas and wanted to send me something to compensate me for any expenses I had because of her stay. Adding and subtracting, I wound up $2 in the black.

A much more striking example of my third principle came true in a situation concerning my first two contracts. I spotted a solicitation from the FDA to perform leak testing on cardiac pacemakers. I knew exactly how to approach this problem, and it fit in well with the capabilities of my garage laboratory. However, when I phoned the FDA program manager, he told me that I was out of sync with the funding cycle, and that they did not plan on awarding that contract that year. Since I had very little to do in my lab at the time, I made him a deal that if he sent me the equipment to do the job, I would do it for free, in hopes of having an edge to bid the contract at a later date. To my astonishment, several Federal Express boxes of equipment appeared on my doorstep the next day, containing almost all the equipment I needed to develop the testing technique, including a set of calibrated leaks. I went to work right away.

In the meantime, I formed a collaboration with another scientist who had a special holographic camera I needed to bid my first U.S. Department of Defense (DoD) contract on nondestructive testing. We won the contract, but the day before negotiations, this scientist dropped out of the collaboration, leaving me stranded with the immediate need to buy or lease a $15,000 camera in order to close the negotiations on my first contract. I was almost completely broke.

Remember those Federal Express boxes of equipment? They contained a camera just like the one I had lost the use of. I was able to successfully complete both the DoD and the National Institutes of Health (NIH) projects.

Obviously, many other principles are essential to running a successful company. Have patience. Go with the flow. Roll with the punches. Learn to laugh at yourself. Notice you are being stabbed in the back, but don't let them know you know. It's very important to remember where you came from. Remember that there were many people who truly believed you could not make it. Remember that there were many people who always knew you would. Remember how valuable it was when someone finally listened to your idea. Remember what it was like to struggle with who you were and where you were going. Remember all this, and be kind to those around you, since you really don't know what the future holds for anyone, nor whose idea will be the ultimate breakthrough. And always keep in mind that arrogance never made a success of anyone.

Of course, there is the important aspect of managing the personnel at your company. Running a high-tech company requires much more than technical expertise and knowledge of the mechanics of negotiation and finance. It also requires managing intelligent, sometimes difficult personalities, both male and female. When I say male and female, I do not mean two categories of people with different physical characteristics. Rather, I am referring to two categories of emotional, intellectual, and mental traits that sometimes exist in the same person.

9.4 Conclusion

Is running my own company hard? Yes, in fact, it is. It is like raising a child. The only way you can learn how to do it is to do it. Many people give you advice from the outside, saying what they would do if it were their business (their child). But it is a lot easier to give advice from the outside than to actually do it yourself. You just can't read a book and be an expert on running a small business. You have to do it.

Is running my own company hard? Yes, in fact, it is. Would I do anything else in the world? No, in fact, I would not.

10

A Case from Russia: IPG Photonics

Valentin Gapontsev
Founder
IPG Photonics

I was born in Moscow in the early days of the Second World War, when my country's very existence was in great jeopardy. In 1941, my family was evacuated to Ural, where we were rescued by German peasants who had been exiled to Ural as enemies of the Soviet authorities. My father was an artillery captain who joined the military in Byelorussia in the first days of the war. He became a commander of one of the first partisan detachments there and finished the war in Berlin as a colonel. We reunited with my father in 1944, after he rejoined Soviet troops following a year and a half in a concentration camp. In 1946, we moved to Lvov in western Ukraine, where I spent the next 20 years. I learned from my father the importance of being strong in the face of adversity. I can think of no more important lesson for someone starting a business today.

10.1 Laser Opportunities

I first became interested in science in school and went on to receive a master's degree in electronics from Lvov Polytechnic Institute. After spending three years working at a high-tech company in Lvov, I returned to receive my PhD at the Moscow Institute of Physics & Technology. My first real taste of R&D as a career was in developing systems for a Soviet satellite program.

Although I was able to use my background in electronics, the more time I spent in this first job, the more I felt the need to move on to something more exciting. I then had the good fortune to get involved with the Soviet Academy of Sciences (SAS).

When I arrived, the SAS was an impressive network of science institutions, rather like a cross between Bell Labs, Caltech, MIT, and NASA. It was an amazing organization, home to some of the world's most outstanding scientists and Nobel laureates. This was a place that welcomed me into the field that would become my life's work: the fundamental physics of light and laser technology. The SAS was also a place of freedom and autonomy from tough bureaucratic regulations.

I started at the SAS in 1964 as a post-graduate student, working hard to strengthen my knowledge of basic physics and the laser science industry. In 1972,

I received a PhD in laser material science, concentrating on the development of new solid-state laser materials and research of nonradiative processes of multiphonon relaxation and energy transfer between rare earth and transition metal sites in laser glasses and crystals. I became known throughout the world as an expert in the field, publishing more than 200 papers and, with my colleagues, receiving numerous Soviet and international patents.

My applications included the first industry-grade neodymium and erbium-doped phosphate glasses, eye-safe rangefinders, precise intra-body optical temperature meters for hyperthermy methods of cancer treatment, and working in high electromagnetic fields, among others.

However, after more than 25 years, I began to see declining opportunities at the Russian Academy of Science (RAS), the successor to SAS. The RAS had become overcrowded, and research funding was progressively more difficult to obtain for working scientists. In addition, substandard equipment, especially for experimental physics, caused a major technological gap compared with Western research centers. As a result, the scientific yield per scientist at RAS was increasingly falling behind our Western colleagues. Another significant problem was the very low quality of life for Russian scientists who had shunned administrative careers. To me, there was no sense in spending the rest of my life in an organization with such inefficient research efforts.

In spite of my dissatisfaction, I probably would still be there today had it not been for the political upheaval and new sense of freedom that was sweeping across the Soviet Union in the late 1980s. Seeing the fall of the Soviet Union and the birth of the new nation of Russia with Western-style political liberties inspired me to leave the RAS bureaucracy. With the few thousand dollars I had saved, in 1990 I set off on my own and started the private enterprise that today is IPG Photonics.

10.2 Fiber Emerges

Starting a business gave me independence, but it also presented tremendous risks. After a career 100% focused on laser material physics, I had never even managed a business, much less started one from scratch. Unlike many other technology companies emerging in Europe and America at the time, my business did not have financial support from government contracts or grants. In addition, conventional wisdom is that successful entrepreneurs start young, but I was 52. Most important, the political disorder in Russia led to a kind of economic chaos that was far better in generating threats instead of opportunities for small business.

Betting everything I owned on the success of my company, I really had no choice but to succeed despite the highly challenging environment. IPG started in the basement of a small laboratory in the Institute of Radio Engineering (IRE) in Fryazino, near Moscow. The team included only a few students. My partner, Dr. Alex Schestakov from the Industrial Laser Institute (ILI), joined me with a group of laser specialists. In the beginning, we made and sold customized glass and crystal lasers, wireless temperature meters for hyperthermy, and laser components.

In parallel, I began to develop new high-power lasers based on fiber optics. During that time, fiber lasers became very popular in Western scientific circles, but the achievable power didn't exceed 10 mW. Moreover, nobody could imagine that a micron-size fiber could emit more power. I was the first to forecast and demonstrate experimentally that it was possible to make 10 W fiber lasers.

Working night and day, and on a shoestring budget with just a handful of close associates, we quickly made the first prototypes. It became clear that fiber lasers could become a major development, and we had a great chance for success. On the other hand, our intensive efforts toward marketing more conventional solid-state and glass lasers didn't give us much commercial success. The Russian market was completely frozen, and the Western market did not demonstrate any opportunity for similar technologies. I realized that the only chance for a Russian start-up with no financial support to succeed in the world market was to introduce a new, disruptive technology. That technology was high-power fiber lasers. In 1992, I made the decision to focus all my limited resources in that direction. Alex's team didn't accept my proposal and we parted ways, but we have remained friends.

Finally, good fortune struck, and we won our first research contract from the large Italian telecommunications carrier Italtel to develop our first marketable product, a high-power fiber amplifier. It didn't take long after that for us to win two more research contracts from the same customer. In total, the three orders were valued at $750,000—a lot of money for a small Russian company.

Following this success, we developed erbium fiber amplifiers, which used innovative pump schematics and fiber solution. This technology allowed entry-scale power into hundreds of milliwatts that were 10 times higher than others had achieved at the time. Italtel wanted to introduce our technology in the market immediately, but their business model couldn't accept the risk of working with a small supplier from Russia. So, they convinced me to transfer the component production to Italy. I proposed opening an IPG subsidiary in Europe, and they agreed.

10.3 Worldwide Growth

In 1994, it was not easy for a Russian scientist to walk into another country and build a new manufacturing operation, but with perseverance—and some luck—we succeeded. We met another serious customer in Germany, DaimlerBenz Aerospace (Dornier branch). They needed a compact, efficient, eye-safe laser transmitter for a helicopter obstacle-warning system, quickly. All previous attempts to make the transmitter utilizing conventional lasers proved unsuccessful. I proposed a new fiber solution, and they agreed to fund it if it was developed in Germany. As a result, we opened a second location in Berlin. One year later we bought a small facility in Burbach, near Frankfurt, where IPG built a high-grade research and manufacturing plant.

During the next few years, we raised IPG's reputation as a highly regarded engineering company and a pioneer in advanced high-power fiber lasers and amplifiers. We developed hundreds of unique products for various applications, many of which did not have competition until now. We sold them at an annual run rate of about $5 million to customers in Japan and the United States as well as in Germany and elsewhere in Europe. However, it was clear to me that the company could not grow without mass production. In 1997, we registered our first large OEM customer sale in fiber amplifiers. The contract was for high-power, multi-port amplifiers from Reltec Communications, a manufacturer for broadband fiber-to-the-home systems being deployed by BellSouth in the United States.

The Reltec contract vaulted IPG to the status of a high-quality OEM supplier and, most important, transformed us into a global telecom active component manufacturing company. To satisfy demand from Reltec and BellSouth, and to serve a growing number of U.S.-based customers, we opened manufacturing facilities in Italy and the United States in 1998.

By 2000, IPG had grown into a profitable $52-million company. Our customer list included Alcatel, Fujitsu, Lucent, and Siemens along with Marconi (Reltec). We had just raised $100 million in exchange for less than 10% of IPG's equity from a consortium of investors consisting of Merrill Lynch, TA Associates, APAX Partners, Robertson Stevens, and Marconi. Our business was growing rapidly, and we were preparing to go public.

Not even I would have believed back in 1990 that it would take only nine short years for the head of a small physics research lab in Moscow to become the CEO of a leading telecom technology company with worldwide manufacturing and distribution operations.

In spite of great success in the telecom arena, we developed and distributed a variety of other fiber lasers for diverse applications in additional markets. For example, we started to ship our new pulsed ytterbium fiber lasers in large volume to SUNX, one of the leaders in the Japanese marking market. In addition, promising achievements were being made in the free space and satellite communications market, where our high-power transmitters were recognized by industry leaders as the best in the world. During this period we received exciting results in the development of our present flagship technology—the multi-kilowatt diode-pumped fiber lasers.

10.4 Risk and Rewards

However, it didn't take long for things to change for the worse. By the end of 2000, telecom capital spending had evaporated. Most of the large- and mid-sized telecom hardware and components manufacturers lost 70% to 90% of their business; many others shut their doors. Our revenue was down nearly 60%, and yet our suppliers weren't budging on pricing or terms.

With our future on the line, I thought back to earlier times, when following the principles of freedom and self-determination had pulled us through. When other companies in our market, even cash-rich companies, had frozen investments and started to cut their staff, we made the critical decision to invest in our future. We invested nearly all of our remaining capital in the development of advanced high-power products, advanced mass production lines and, most important, in a high-volume production facility to make our own high-power pump diodes. This new facility enabled us to cut our dependence on sole-source suppliers and radically cut production costs.

In making this investment, we were effectively betting the company on the competitive benefits of vertical integration. If we could manage the price, quality, and quantity of our components, I knew we could accomplish this goal. We equipped IPG with the capability to manufacture every crucial component within our laser and amplifier products.

We manufactured everything from the diodes, which pump the amplifiers and lasers, to the various specialty optical fibers that generate and transmit the laser emission, as well as many other optical and optoelectronic components. We believed that with a vertically integrated supply chain, we could produce higher-quality diodes for up to 90% less than we paid our suppliers.

After only three years, we developed world-class pump diodes that exceeded the power, brightness, and reliability of those produced by the leading supplier. But the main benefit was our ability to dramatically reduce manufacturing costs. Moreover, IPG built and deployed the world's largest pump laser diode production line and manufactured more diodes than all our competitors put together.

The availability of high-quality, low-cost pump diodes opened up an opportunity for IPG in 2003. We introduced to the market a revolutionary new generation of super-power, multi-kilowatt fiber lasers that, in a short time, completely changed the competitive landscape in the metal cutting and welding market segment by displacing conventional crystal and carbon gas lasers.

Now IPG's fiber lasers provide the automotive, aerospace, heavy, and other industries the highest power, beam quality, overall efficiency, and reliability at lower cost per watt. Since they are much more compact, lighter, maintenance free, and user friendly, their introduction into the market was accompanied by much higher standards in performance, reliability, cost, and service.

Armed with this new unique range of super-power products and competitive cost and quality advantages, we have captured a steadily increasing share of the worldwide market.

At the same time, working closely with our customers, we have spurred the emergence of major new markets for lasers. After hitting a low point of $22 million in sales in 2002, we have grown to more than $143 million in 2006. During the past four years, our compounded annual growth rate (CAGR) has exceeded 55%, and we are very optimistic about our future based on current market trends.

Looking back at the past 15 years, I would do very few things differently. Although I may have wanted to become an entrepreneur earlier in my career, it would not have been possible in the Soviet Union of that era. With hindsight, we would have been better off if I had reacted more quickly to the telecom meltdown, but I was not alone in thinking the downturn would be brief and shallow.

Looking forward, my dream is to see lasers—like computers—become a tool of choice in mass production, rather than being viewed as a last resort in many applications. I intend for IPG Photonics to play a pivotal role in realizing this dream, and being sure to maintain our independence along the way.

10.5 Some Final Comments

IPG Photonics was founded in 1990 as a small optics company in the Soviet Union. Since that time, IPG has become a leading developer and manufacturer of high-performance fiber lasers and amplifiers for diverse applications and markets, reporting revenues of $143.2 million in FY2006. The company now operates manufacturing facilities in the United States, Germany, Russia, and Italy, as well as sales offices in Japan, China, Korea, India, and the United Kingdom. Major developments in IPG's history include moving to a vertically integrated manufacturing model in 2000 and going public at the end of 2006.

As a company dedicated to controlling its own destiny, the decision to move forward with a public equity offering was not easy for us. By 2006, however, the potential benefits of an IPO were too compelling to ignore.

Reports from the marketplace told us that it would be hard to increase our penetration much further without providing customers with the financial transparency and broader awareness that publicly held companies enjoy. Going public also was the most reasonable way to provide our long-term investors with the liquidity they deserved.

While we experienced some small changes after becoming a public company, in spirit we are still very much the company I founded in a small lab in Moscow, a company dedicated to scientific discovery and technology innovation and a company willing to risk anything to preserve the freedom we cherish.

11

Wacko WYKO

James C. Wyant
Cofounder and former President of WYKO Corporation
Professor, University of Arizona
Tucson, Arizona, United States

The WYKO Corporation was founded on Dec. 27, 1982, to design, manufacture, sell, and service metrology instruments for many applications, with the largest market being in the magnetic data-storage industry. In the next few pages I will discuss the formation, growth, and eventual selling of the WYKO Corporation to Veeco in 1997. For me, this was an unbelievable experience that was more fun than I ever dreamed anything could be.

11.1 Technical Innovative Years

WYKO grew out of the research my students and I did at the Optical Sciences Center at the University of Arizona, but it actually began at the Itek Corporation where I worked after getting my PhD in optics at the University of Rochester in 1968. For a few years, my job at Itek was nearly ideal. We had freedom to do whatever we wanted to do, we had plenty of money for equipment, and we had very smart people to work with. Most important, we had marketing people who brought us interesting problems we could work on if we wanted to. One problem they brought us concerned correcting for aberrations in optical systems and correcting for atmospheric turbulence so we could obtain better images of Russian satellites. This led us to develop what is now called adaptive optics (then called active optics). In the process, I developed ideas for what we now call phase-shifting interferometry. For the purpose of this story, interferometry is a good way of getting interferogram data into a computer. It turns out that phase-shifting interferometry had been invented earlier by Carré, but it was several years later that I became aware of Carré's French paper. At the time, we built some adaptive optics systems, but they were very primitive because we did not have the solid state detector arrays, computers, and electronics that later became so important for high-quality adaptive optics systems. Phase-shifting interferometry could not be implemented well with the technology present at the time, and it was essentially useless, so I nearly forgot about it for several years.

11.2 The Product Idea

In 1974, I joined the faculty of the Optical Sciences Center at the University of Arizona and began building a research group. About seven years later, I visited the Union Carbide Y-12 plant in Oak Ridge, Tennessee, where they were making diamond-turned mirrors and parts for the atomic bomb. I saw they were using several interference microscopes to determine the surface finish of the diamond-turned parts. They took Polaroid pictures of the interference fringes, then several people analyzed the interferograms using a ruler and pencil to determine how straight the interference fringes were. I figured there had to be a better way of determining the surface finish.

Figure 11.1 First (almost) phase-shifting interference microscope.

When I returned to Arizona, I contacted two friends at Los Alamos Scientific Labs, Tom Stratton and Walt Reichelt, to see if they would fund the development of a better way of measuring the surface finish of the diamond-turned mirrors they were having made for their laser fusion system. They almost immediately gave me funding. I found a brilliant student, Chris Koliopoulos, to work on the project for his PhD dissertation. By this time, some Reticon detector arrays were available, and the Z80 microprocessor had just come on the market, so after a little fumbling and borrowing some interference microscope parts, we were able to put together a crude phase-shifting interference microscope system. Photos of the first system do not exist, but Figure 11.1 shows a photo of a system similar to that first one. We thought it was fantastic! The only problem was that the measurement results were complete garbage. The system was essentially a random number generator, but we were so excited about it we didn't care. While we knew optics fairly well, we did not know electronics, and we were not good at interfacing the detector to the computer. We took it to one Optical Society of America (OSA) meeting and showed it to a couple hundred people. Only one person realized the results were garbage. I guess what was saving us was that if you measure surface roughness, you get results that look rather random. One company approached us about buying the idea, and we offered to sell it to them for $10,000. They agreed, but after they studied the system more, they realized our measurements were useless, so they backed out of the deal. They did not realize that, while our implementation was not good, our basic idea was excellent. We were so lucky!

We continued to work on the system at the university, but we were having trouble getting good data for any period of time. Then we had two gifts from heaven. The first gift was actually our first of many gifts from IBM. The upper management at IBM felt there must be some technology at the University of Arizona that would be of use to them, so they sent a wonderful engineer, Bharat Bhushan, to the Optical Sciences Center to see if we had any technology to help his work. Bharat was involved in measuring the surface roughness of magnetic tape and asked me if we could measure the tape for him. I told him we could and described the phase-shifting interference microscope system we had. I forgot to mention that the measurement results were generally useless and looked more like random numbers than surface roughness.

He gave us some tape samples to measure, and we measured them for him. Fortunately, the surface roughness of magnetic tape was pretty random, and by some miracle our measurements correlated somewhat with some other measurements he had done, so he became very excited about our system and nearly worked us to death doing measurements for him. He once gave me some samples at 10:00 PM Sunday and said he needed the results Monday morning. He got them.

He then decided he wanted to buy a system. We told him we did not want to build another system at the university, but we had started a company called WYKO (WY from my name and KO from Chris Koliopoulos' last name) and that we would sell him a system for $100,000, but we needed $60,000 up front to buy the parts. He said OK. (Note that we should have worried about the university owning the IP and gotten permission from the university to sell the system through our company, but in 1982 we didn't worry about such things.)

The second gift from heaven was a letter from a recent PhD graduate from Sheffield University in England, Keith Prettyjohns, who wanted a post-doc position. The great thing was that he knew electronics and had done some work interfacing Reticon solid-state detector arrays to computers. I quickly gave him a post-doc position, and we had someone who could help us make a phase-shifting interference microscope system that would actually work. It turned out that Keith was fantastic.

11.3 Founding WYKO

In December 1982, we received a $60,000 check from IBM, the down payment for our first sale. I rushed off to the bank to cash the check when I learned we had a problem. I had told Bharat that we had formed a company, WYKO Optical, Inc., but we had never actually formed the company. The check was made out to WYKO Optical, Inc., and we could not cash the check.

We formally started WYKO on Dec. 27, 1982, so we could cash the check. The founders were Chris Koliopoulos, who by then had completed his PhD and was an assistant professor at the Optical Sciences Center, Keith Prettyjohns, my post-doc, Steve Lange, an optical researcher who worked on my projects at the university, and me. The original goal was that Steve would leave the university, run WYKO, and make us all rich, and the rest of us would stay at the university.

We rented a small office off campus, and our WYKO goals were to have fun and make money. Unfortunately, neither was happening. We were not having fun, and money was not flowing in through the door. The problem was that we could not make the IBM system work. We were able to make simpler systems for other customers, so we had some income, but the system using the two-dimensional detector array that we sold to IBM would not work well enough for us to deliver it to IBM.

Every month, I would send a letter to the IBM purchasing person telling him that we were unable to ship that month, but we would ship the next month. The next month, I would change the date on the letter and send it again. IBM was extremely nice; they gave us more time and did not ask us to return the deposit.

Finally, in September 1984, I knew that things had to change. WYKO could not continue this way. We made several changes, one of which was that I stopped doing research and went to 20% time at the university, going to WYKO full time, or at least 80 hours per week. I thought this change would be for two to five years, but it continued for 13 years. John Hayes, who had just finished his PhD working for me, also joined WYKO at that time, and he was excellent at designing real products.

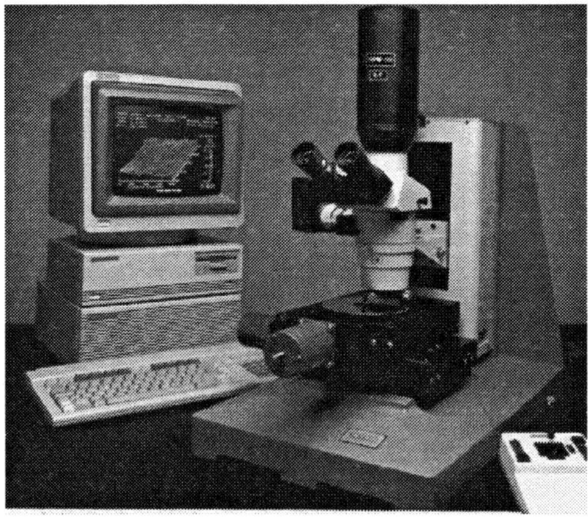

Figure 11.2 TOPO-3D phase-shifting interference microscope.

11.4 Growth Time

WYKO changed rapidly. We were able to deliver the system to IBM and develop an extremely successful phase-shifting-interference microscope, TOPO-3D, shown in Figure 11.2. I found that by putting my full effort into WYKO I was able to get others to put their full effort in as well. People will work extremely hard for you as long as you work as hard as you ask them to work.

By then we had moved to a larger second location. We shared this second location with a dentist and a podiatrist. Since our name was WYKO Optical, Inc., we had a lot of people stopping in to buy eyeglasses. We decided to change the name of the company to WYKO Corporation.

Sales and profits were excellent, so we developed more products and moved to our third location shown in Figure 11.3, a photo taken by a visitor from Kodak. I thought the rainbow in the photo was fantastic, and I felt we had found the pot of gold at the end of the rainbow. This location was across the street from the university, so I could easily walk to the university to teach a class, and students could easily work for us. University students make great employees. They are smart and they work hard, and they do not demand high wages. It gives you a great opportunity to learn more about the students, and you can easily hire the best after they graduate.

Figure 11.3 The third WYKO location.

During this time we had a big surprise. We started selling a phase-shifting Fizeau interferometer for the testing of optical components and optical systems, and one day, without any warning, we received a letter telling us we were being sued for patent infringement. There was no advance warning that if we did not stop selling we would be sued, but we were simply sued for infringing three patents.

Rather quickly, two of the three patents were removed from the lawsuit, but the suit over the third patent lasted for 10 years. We had been careless. We had known about the patent under question long before we introduced our product. While we thought interferometry experts were well aware of everything in the patent long before the patent application was filed, we discussed the patent with an expert patent attorney before we introduced our product. He told us what we had to do to avoid problems with the patent, and we followed the attorney's advice, but the smart thing would have been to construct things completely different from the patent, which we could easily have done.

While the patent lawsuit was a terrible experience at the time, in hindsight it helped us, and me especially. It made me realize that a patent does not have to be a Nobel Prize-winning idea. Patents about rather ordinary things can be very important. We began filing our own patents, and by the time we got ready to sell the company, we had some 45 patents that made the company much more valuable. Also, because of the patent lawsuit, I stayed at the company for several more years than I had initially planned to. During this time, I had a lot of fun and the value of the company increased tremendously.

Figure 11.4 Phase-shifting interference microscope for measuring the inside of an engine bore.

Figure 11.5 Solder bump measurement system.

As the business grew well, we purchased a 110,000 sq. ft. building that had previously been occupied by IBM. We also hired several previous IBM employees, including Esther Davenport, who had an MBA from Northwestern University and added much business expertise. We added a former IBM executive, Carmon Rosato, to our board of directors, and he gave us a lot of valuable advice. We won several awards such as the SPIE Technology Achievement Award, R&D 100 Award, and several Photonics Spectra Awards.

While measuring magnetic media remained our largest market, we added instruments for the automotive and semiconductor industries. (See Figures 11.4 and 11.5). An enjoyable product, but a small financial disaster, was the foot scanner shown in Figure 11.6 orthotic. The idea was to replace the plaster casts podiatrists use to measure the shape of your feet for making orthotics. Our foot scanner used a phase-shifting technique to rapidly measure a patient's foot, and then this data could be sent over a phone line (the Internet was not yet popular) to a lab that could mill an orthotic. Technically, our foot scanner was fantastic, but we wanted to sell the product for $12,000, and the podiatrists wanted to pay $2500 for it. Thus, we did not sell many of the systems. It was a really fun project and well worth the $1 million or so that we lost on the project. (I might add that I still have a couple of the foot scanners in my garage, and if any reader is interested in purchasing one, I can give you a real deal.)

11.5 Cashing out

By the time 1996 came along, I was 53 and knew it was time to sell WYKO. While I was still having fun, I wanted to go back to the university full time before I became too old—and I was tired. I had been working essentially seven days a week since 1984, and I needed a rest. We talked with four companies about buying WYKO, and we finally decided to sell to Veeco. Essentially, every manufacturer of hard-disk drives in the world was using our equipment for evaluating hard disks and recording heads and using Veeco's process equipment for manufacturing the disks and heads, so it seemed like a good match. We worked out a deal where we would trade WYKO stock for Veeco stock in a tax-free swap.

Figure 11.6 Foot scanner.

The $60,000 IBM gave us as the down payment on the first system sold was used to fund the WYKO start-up, and we never had to go to outside investors. Thus, we owned the company, and we did not have to split the profits with investors. We were so lucky!

In July 1997, we completed the deal, and I went back to the university. After sleeping and playing with Mathematica for a few months, I once again became a full-time professor on Jan. 1, 1999, director of the Optical Sciences Center. In July 2005, the Optical Sciences Center became the College of Optical Sciences, and I became its first dean. Once again, I was lucky, and my timing was nearly perfect.

11.6 Biggest Surprises

Involvement with WYKO's start-up and management taught me a lot. My four biggest surprises were:

1. **The difference between research project and product.**
   At the university, we try to get an instrument working well enough that we can take a few measurements and write a paper. If you are selling a product, you must have one that you can ship halfway around the world and take out of a box. An unskilled operator must be able to easily operate it, and the instrument must continue to work well for a long time. This is very difficult. WYKO has caused me to have much less respect for many of the papers I read.

2. **The difference between the cost of parts for a product and the cost to design, sell, produce, and service the product.**
Sometimes I hear a person say the parts cost only $20,000, therefore selling the product for $35,000 will result in a lot of money. The company would, of course, go bankrupt with such a small markup. I feel the selling price must be high enough that if later the customer has complaints about the instrument, we do not resent fixing the problem.

3. **The amount of money required to run the company.**
A product has to be conceived and developed. Prototypes have to be made, and a final product has to be constructed and evaluated. The instrument has to be marketed and sold. A product has to be delivered to a customer, and then the manufacturer does not get paid for at least another 30 days. A lot of money has to be spent before any revenue is realized.

4. **The number of personnel problems.**
While WYKO's turnover rate was much lower than industry norms, and we certainly hired wonderful employees, the personnel problems seem to grow faster than the number of employees. Hiring mistakes are inevitable, and the best policy is to correct hiring mistakes as soon as they are recognized. When personnel problems upraise, they should be solved before they become worse.

11.7 Most Important Factors

The most important factors are to:

- Have a product that people want to buy. When you are thinking about introducing a new product, first ask, "Does it work?" If the answer is yes, then ask, "Does anyone care?"
- Be willing to do whatever it takes to get a job done.
- Be good at hiring people. This is extremely important because it is always the employees that make the company succeed.

I think it is essential that a company leader must have absolute dedication to the company. Timing is extremely important and probably the most important item is good luck.

Starting and growing a company is not for everyone, but for the right people it can be an extremely rewarding experience and I strongly recommend it.

12

The Ocean Optics Story in a Nutshell

Mike Morris
Cofounder and President of Ocean Optics Inc.
United States

In this chapter the path to success followed by Ocean Optics is presented. Comments concerning the things to do first, on the business plan, on sales and marketing, on the key elements to the sales platform, on the role of trade shows and conventions, are addressed. An extensive due diligence checklist, very useful for entrepreneurs, is included as an appendix.

12.1 Introduction

The story of Ocean Optics is at once both unique and familiar. We started in an entirely different industry, making and selling pH testers for aquariums, as the original garage-based business. It wasn't even a full garage, just a 10'x10' aluminum shed in my back yard. In 1989, we made the leap to high-tech by winning a Small Business Innovation Research grant from the U.S. Department of Energy (DOE). The idea was to put our aquarium pH test material at the end of an optical fiber and use it in oceanographic research. After we quit our day jobs and started work on the fiber-optic pH sensor, we discovered that a key part of the proposed instrument didn't exist. We had assumed you could buy a small spectrometer to fit into an underwater housing. Suddenly, we were faced with the necessity of inventing one. For two years we labored in R&D, and in 1992 we launched our first commercial product.

No, it wasn't the fiber-optic pH sensor; it was the world's first miniature spectrometer. This spin-off technology became the core product in a family of fiber-optic accessories and gadgets that grew at the astounding rate of 1.4 new products per week. The growth in accessories was driven by the growth in applications that our customers dreamed up.

As all successful growing companies discover, when things are going well, you will run out of cash. We raised about $1 million in cash by selling stock in two private placements. The investors were our friends, families, colleagues, and distributors. Because the investors were all friends, we were able to keep the legal costs very low.

Along the way, we were lured by the big money that was being made in the telecom mania. We tried our hand at what's called a roll-up strategy. The plan was to borrow money, buy some small companies, bundle them into a bigger package, and sell this to Wall Street in a public offering. We actually did two acquisitions and started a new division before the rug was yanked out from beneath us. In 2000, the telecom bubble burst, Bank of America called our loan, and we were facing insolvency.

We regrouped and worked very hard to raise cash by selling off assets. We sold nonperforming portions of the business and emerged stronger and more profitable. In 2004, we successfully sold the company to Halma p.l.c., a UK-based public company that specializes in acquiring and mentoring smaller engineering-based companies.

Our investors were very pleased. They received about a 50:1 payout for their 10-year investment. The employees were pleased because in Halma we found a company with resources to fund our dreams along with a hands-off management style that let us have our own dreams.

Today, Ocean Optics employs about 250 people in six facilities around the world. We have sold our first invention, the world's first miniature spectrometer, to more than 75,000 customers. Our spectrometers have flown on the MIR Space Station and the space shuttle and will be on board the next Mars Rover. Our spectrometer is used to monitor patients' blood oxygen in critical-care facilities, measure the color of LEDs, monitor wavelengths and mode structure in lasers, detect explosives and drugs at airports, and teach students the fundamentals of spectroscopy in college and high school teaching labs.

It has been a great deal of fun.

12.2 Creating Wealth

Robert Kiyosaki did a great job explaining how to make money with his wealth-creation quadrant. Most of us were taught by our parents and teachers to become good employees and trade our labor for pay. Some of us pursued additional education to become professionals (engineers, scientists, lawyers, and accountants) and trade our expertise for pay. The problem with these strategies is that there is no accumulation of equity, no residual income. Once you stop working, you stop getting paid. In the two other quadrants—industrialists and investors—wealth is created by leveraging people and money, respectively, to build wealth.

Being a smart investor is an essential ingredient for everyone's personal financial strategy, but being a full-time investor requires having enough starting capital. Being an industrialist can also be very lucrative and, as with investing, start-up capital is pretty handy. In both cases, not having capital is just an obstacle to overcome, and having capital doesn't guarantee success.

The benefit of owning a business is that you can grow the value of your company. The retained value or equity is something you must recognize and nurture carefully. When you make day-to-day business decisions, you must always ask, "Does this add to the equity of the company?" If the answer is no, or worse, if the activity will diminish the equity, then don't do it.

Equity takes on many forms. Some are obvious, like intellectual property. If you have a patent, you can sell it or license it to others, you can borrow money against it, and you can show its value on your balance sheet. Others are not so obvious to the new entrepreneur, but intangible equity will be the most important component in the end.

The most important part of Ocean Optics' equity was not its technology but its business model. We created a system for developing, manufacturing, and selling products that generated profits and cash flow. When we sold the company, the price we received was largely based on a multiple of profits that the business generated. Our tangible assets were just a small fraction of that price. This system or business model has many components, including our brand name, our advertising, our products, our distributors, and of course, our customers. While many competitors duplicate certain elements of what we do, it's the whole package that makes Ocean Optics unique and that gives the company its value.

While wealth isn't everything, it sure comes in handy when you try to buy things. Its also lets you fulfill your charitable and altruistic impulses. I'm very proud of the success of Ocean Optics, mainly because we were able to create six millionaires and a dozen or so half-millionaires among our investors. If we had the means to calculate the total economic impact of Ocean Optics, it would be huge.

12.3 Last Things First

"If you don't know where you are going, you may end up somewhere else" is something that baseball legend Yogi Berra might have said. It's very useful to have an end-game or exit strategy in mind when you start your business. If you try to raise money, it will be essential. There are two main options: going public (selling the company to the public) or selling to another company. There are other less common alternatives, like employee buyouts and such, but they are harder to engineer.

I like to visualize, so picture yourself selling your company in 10 years. How will you do it? Who will you approach? Why will someone buy your company? When you sell a company, the buyer will go through something called due diligence. It's a discovery process that looks for reasons not to do the deal. You will need to create a due diligence file containing all the stuff requested in the due-diligence checklist (see the Appendix for an example). There are obvious things like financials, but also included are not-so-obvious things like minutes of all your board meetings, environmental reports for any property that you owned, copies of all contracts (including purchase orders), and lists of all potential liabilities. You should make a due-diligence file now and keep it up to date. It's a lot easier than trying to recreate it later.

One last point: keep your business squeaky clean and very simple. Legal entanglements lower the value of your company.

In the end, someone will buy your company either for strategic reasons (access to technology, markets, etc.) or for the cash flow. There are only a handful of potential strategic buyers, so start courting them early and keep them up to date with your progress. Make sure you understand their strategy, so you can optimize your strategic value to them. There are many potential cash-flow buyers, so focus on having good cash flow! Keep track of who is doing acquisitions in your industry and make sure they know about you and your company.

12.4 Business Plan

Don't worry about writing a business plan; do worry about planning your business. We spent more than a year planning our business while we also worked on our SBIR research project. We were very fortunate to have gotten free mentoring from Dawnbreaker, a consultant hired by the U.S. Department of Energy to help their SBIR companies commercialize their research. We went through a series of exercises and homework assignments that made us think. That's the key; often it's easy to get too busy to think. You need to slow down and give it a try! While most people think of the business plan as a way to get investors, it's much more valuable as a blueprint for success. At Ocean Optics, our original plan served as a daily guide for more than 10 years, giving us a framework for thinking about the thousands of issues that arise in the course of running a business. In actuality, we never raised any money with the written document. Our first investors believed in us as individuals. Our one outside investor was a venture firm that didn't read the plan at all. They just happened to have a few hundred thousand dollars left over and needed a nice, quick, small investment to round out their portfolio.

12.5 Sales and Marketing

To mangle a famous quote, "Sales is not the most important thing, it's the only thing!"

Ocean Optics has always been first and foremost a sales and marketing organization. We view research and development as a sales and marketing activity, and we view sales and marketing as an experimental science. So, in a practical sense, what does this mean?

We designed our products by first imagining how we would sell them. I mean we actually pictured ourselves talking to a prospect, discovering their needs and wants, offering a solution and getting paid to do it. That is the essence of being in business. If you can't do that successfully, you won't have a business.

Early in my career, I sold laboratory products—chemicals, glassware, instruments, and such. Despite having a 10-pound catalog to haul around (this was long before the Internet or even PCs), I often found myself investing time to understand a prospect's problems only to realize we didn't have a product to solve that problem. It was frustrating and, of course, inefficient. No solution, no sales, no commission.

At Ocean Optics we decided to eliminate this bottleneck by emphasizing a particular attribute in everything we do—flexibility. Our products are highly modular, all the modules can work with each other, and all the modules connect with flexible optical fibers. Trillions of combinations are possible; all we have to do as salespeople is pick the best combination for a particular customer. Our company also has a flexible attitude. We are always willing to try new things and eager to work with customers. We are not too scared of failure to try. In practice, we actually develop new products and applications during a sales call. We don't worry about market size, market niche, core competency, return on investment, or any of the hundreds of other usual business school things that get in the way of creativity.

Here is a typical Ocean Optics sales call:

A prospect calls and asks, "Can your spectrometer measure <stuff>?"

We say, "I don't know, what's <stuff> and why on earth do you want to measure it?"

An hour later we have developed a new way to measure *<stuff>*, we have gotten an order for the first system, and we have a vague concept for a new *<stuff gadget>* accessory we need to develop in the next few weeks to fill the order. We'll make a few *<stuff gadgets>*, add them to our product line, and then through our experimental marketing we'll discover a few more uses for them. In our 13-year history, we have averaged 1.4 new products per week. The only thing keeping us from doing more is lack of brain bandwidth!

12.6 Sales and Marketing: Practical Stuff, Starting on a Shoestring

If you are a multibillion-dollar company, you have resources to do quite sophisticated sales and marketing planning. If you are small, you need to be clever. Here are a few tips.

Know your industry. You are trying to join an industry, so you need to know everything you can about it. Trade journals are free; they list new products, trade shows, conventions and job postings, and have articles about various issues of interest to the industry. They also report on industry leaders, who owns whom, who landed big contracts, who got funding, and who is going public. Read these journals from cover to cover!

Go to trade shows and conventions! If you have a good contact with a company who is exhibiting, you may be able to get in for free. Either way, this is where you meet people and make contacts. All business is fundamentally based on relationships with people. That's why the trade-show industry is doing so well. Face-to-face business is essential. If your business deals with technology, then you must attend professional society meetings. That's where you find out the latest developments and technologies, and that's where you meet people you will one day try to recruit or sell to. After you are in business and can afford it, join the professional and industry associations, attend the meetings, and exhibit at the trade shows!

We made our first sales by walking the floor at PITTCON in New Orleans in 1992. Our prototype miniature spectrometer was in our shirt pocket. We invited potential OEM customers back to our hotel room for demonstrations. It was a very low-budget way to get into the industry. A year later, we had our first trade-show booth at PITTCON in Atlanta. We were so busy we never had a chance to take a break. In four days we did demonstrations and collected leads from 585 prospects.

You need friends! Or, to be more precise, you need to start creating a network of stakeholders, admirers, and advisors. Talk to sales people! Advertising reps are ideal because they talk to everyone in the industry. They will help spread the word about you and help place your new product announcements in return for the promise of future advertising business. They will become stakeholders or people who have a vested interest in your success. Admirers are hard to target; they really just kind of happen. However, when you find you have admirers, treat them well. They make good word-of-mouth advertising. Good advisors are the hardest to find. We found advisors by making the rounds of business development groups like the Small Business Development Center, NASA tech transfer office, and groups at the local universities. We ended up with pro bono legal, accounting, and marketing assistance from firms trying to help grow local high-tech businesses. Eventually, you may need to formalize the relationship by asking some of your advisors to become board members. They may ask for something in return because they will assume some liability for joining the board. At the very least, you will need to offer directors and officers liability insurance.

Our most valuable advice came from a marketing firm. It was valuable because we did just the opposite. They wanted us to change our name to something high-tech like "Spectra Tech" or "Tech Spectra." We noticed the big, impressive booths at trade shows were companies like Hewlett Packard and Perkin Elmer. So, we stuck with Ocean Optics. Be careful, though; when you pick a name or a logo, you might be stuck with it for years to come. Our first logo was a laser shooting at a dolphin. It incorporated the twin ideas of optics and the ocean, but it was a bit sinister. Later, we adapted a clip-art globe. Since 70% of the world is covered with oceans, and all the optics are manufactured on land, it seemed good enough.

We did pick up a very useful concept from an advertising firm that wanted our business. Don't underestimate the value of listening to sales pitches from advertising firms—they always divulge their best stuff to land the contract! We borrowed the idea of a sales platform from just such a pitch. A sales platform is a plan to coordinate all aspects of the marketing message to make sure they are internally consistent. Again, it's the master guide that you use to make everyday decisions about how you talk to people, how you craft your ads, and even what shirts you wear at trade shows. The elements to the sales platform are:

- Audience (people in the target market)
- Objectives (what you want to accomplish)
- Theme
- Storyline
- Image
- Features/benefits/proofs
- Answers to objections

Most sales people are comfortable with the concept of features and benefits, but the longest-lasting messages are in the theme, storyline, and image. Long after people forgot about our 1-nm resolution and high-performance whatever, they remembered our story (we invented the world's first miniature spectrometer because we had to), they felt resonance with the theme "smaller is faster, less expensive and higher tech," and they knew Ocean Optics as the high-tech company with an attitude.

Be different! A principle in ecology states that no two organisms can occupy the same niche at the same time. One will survive and the other will become extinct. So it is with companies and products. Only the different survive. You need to be very clear about what makes your product and your company unique. You can sell differences; you can't sell similarity.

You need to be a professional salesperson. All aspects of your business involve sales—raising money, recruiting talent, and of course, selling products and services. Professionals take charge of their own education, so you need to study salesmanship. There are lots of books, CDs, courses, and coaches. Here is a quick summary of the most important aspects:

- Sales and marketing is communication. In marketing, you are projecting a message and hoping someone listens. In selling, you are asking questions and listening.
- Communication requires error checking. It sounds kind of obvious, but always tell your prospects what you think they just told you.
- Salespeople ask many questions, but one question rules them all: "Why?" If you don't know the why, you can't serve the customer's needs. I told this to a group of fifth-grade students once who were visiting Ocean Optics. One bright kid raised his hand and asked, "Why?" We gave him a prize.

- Sell the vision, negotiate the details. Never discuss minutiae (like price, specifications, delivery) until you have sold the prospect on your concept or vision on how to solve his problem.

If you are a good salesperson, all you really need to make a living is a phone and a list of prospects. That is because it is amazingly inexpensive to design new products, manufacture them, and distribute these goods anywhere in the world. The bottleneck is in getting people to learn about new products and to try them and buy them. The value proposition is in the distribution not of goods, but of knowledge that goes with the products. Applications, tech support, and training are all part of this intellectual distribution model, and the professional salesperson is the intellectual distributor.

Your sales and tech-support people will build one of the most important equity components in your company: market contact. Once you are in a market, you will be exposed to many opportunities. Your contacts will tell you their wants, you'll see their needs, and if you are at least a little clever, you will have ideas for new products or services to solve their problems. So, what's the best way to do market research? Sell a product. Even if it doesn't exist yet, get into the market. Your prospects will quickly let you know what they really want.

Lack of resources is no excuse! Companies never fail because of lack of resources. They only fail because of poor management. I was in debt, getting hounded by bill collectors, and worrying about raising my infant daughters in poverty. I naturally blamed Ronald Regan, the government in general, and other vague entities called "they" for my troubles. If I had said, "Well, I need a million dollars to start a company, and without that money I can't succeed," then guess where I'd be today. Your job as an entrepreneur is to succeed with what you have and to overcome whatever obstacles get in your way. Put things in perspective! If you make only the U.S. minimum wage, you are in the top 8% income bracket in the world. That means you are at 92% of the goal of being the richest person in the world! Stop whining!

Was I lucky? Luck is defined as random, and what every good scientist knows is that randomness averages out over enough data points. So, I suspect I was lucky about half the time. Was I blessed? Absolutely. I was born in the greatest country to two loving parents. I was educated by great teachers at good schools. I went to a world-class university. My fellow citizens made a faith investment in me totaling more than a million dollars, asking only that I try to succeed. I entered an industry that was highly ethical, where many people helped just because they liked us. My friends and family entrusted me with their own hard-earned wealth. The question was never, "How could I succeed?" The question was, "How could I possibly fail?"

13

Experiences Starting a Nano-Sized Company

Richard O. Claus and Jennifer H. Lalli
Cofounders of Nanosonic, Inc.
Blacksburg, Virginia, United States

This discussion summarizes some of the experiences during the creation and early growth of a small high-tech company, NanoSonic Inc. NanoSonic is a spin-off of the Fiber & Electro-Optics Research Center (FEORC) at Virginia Tech. Its first 10 years of operation may suggest several issues related to fiber optics, sensors, and materials.

13.1 Introduction

NanoSonic Inc. was formed by staff of the Fiber & Electro-Optics Research Center (FEORC) at Virginia Tech in 1998, in cooperation with the university. FEORC was created by the Commonwealth of Virginia in 1985 to promote industry/academic interaction in the area of applied optics. The Electro-Optics Product Division of ITT in nearby Roanoke, Virginia, was a primary motivation for its establishment. Over the years, more than 20 small companies were spun off from FEORC by students, staff, and faculty, for a variety of reasons and in a variety of areas related to the central theme of optoelectronics.

NanoSonic's special niche was and remains specialized materials used to form coatings and bulk materials with controlled optical, mechanical, and other properties. NanoSonic's early work built directly on research performed within FEORC. Several FEORC programs concerned the design and production of polymer coatings on optical fibers, either directly during the fiber-draw process or on short fiber segments to implement specific fiber sensor element and system designs. Two particular representative projects concerned fiber coatings with designed modulus in addition to normal cure kinetic properties, and magnetic coatings with designed permeability properties to enable the implementation of magnetic field-directed fiber placement. Both of these problems required the use of a process that could allow the incorporation of multiple macroscopic constitutive properties into a polymer-based material.

The technique that FEORC students and staff developed, demonstrated, and later patented, and that Virginia Tech eventually licensed to NanoSonic, is termed electrostatic self-assembly (ESA). It allows the formation of thin-film coatings of multiple materials—polymers, nanoclusters, biomolecules, and other constituents—conformally onto the surfaces of substrates such as optical fibers. The two interesting advantages of the process for the formation of such fiber coatings were that well-controlled materials could be produced with uniform thickness over the surface of curved substrates, such as optical fibers, and multiple layers of a wide range of materials with very different chemical properties and thermal limits could be co-processed into a single organic/inorganic "nanocomposite," something very difficult to do using vapor deposition or sputtering alone. The difficulty of such coatings also lies in the process. ESA is somewhat similar to the way bones grow, one molecular layer at a time, and, unfortunately, it takes a long time to grow bones. One of NanoSonic's primary objectives has been the transition of the basic ESA method for synthesizing interesting ultra-thin film coatings for research purposes on small substrates, such as optical fibers, to the rapid production of thick coatings or free-standing materials on large substrates or large areas.

13.2 Company Formation

NanoSonic was not formed in order to form a company; it was formed to support a graduate student at FEORC. In 1997, FEORC had arranged a research project involving the fabrication of specialized coatings with a company in the United States; the project would have supported a PhD student, a part-time staff person, materials, and related costs on campus. Unfortunately, the terms and conditions of the proposed contract were not suitable to the university, so FEORC personnel formed a company off-campus to support that PhD student. The company was formed one afternoon by faxing a form and credit card number to a company that provides the service of filling out incorporation forms. Capitalization totaled $1,000; the three initial officers were a physical chemist, an electrical engineer, and a business manager. The first project was performed by the PhD student (the physical chemist) in his kitchen; business administration was handled on the dining-room table of the business manager; and technical and business files were stored in a box in the back seat of a car. Corporate overhead was low.

NanoSonic slowly grew through a number of small contracts, some obtained through the Small Business Innovative Research (SBIR) programs of several U.S. government agencies. Personnel were added gradually, primarily by involving additional graduate students from FEORC who had direct laboratory experience with ESA coatings and analysis methods. Since FEORC had its basis in electrical engineering, most of NanoSonic's initial employees had backgrounds other than chemistry. Gradually a group of chemists joined the company, and they were able to improve and upscale the ESA process and develop new processing methods for the production of related materials. This change in company focus from EE-based fiber-optics problems to ESA process chemistry, was important to the development and current level of success of the company.

13.3 Company Growth and Directions

NanoSonic's growth and direction has been because of the areas of expertise, interest, and energy of its primary technical staff, success in obtaining development programs and material production orders in specific areas, and top-down business planning by the corporation as part of strategic planning efforts. Some of the areas of greatest interest and within the general capabilities of NanoSonic's staff did not lead to successful programs or possible products, while others did. Over several years and building on the experiences of multiple staff, several key product areas emerged. Due to the significant downturn in the optical fiber technology area in the early 2000s, the relative maturity and lack of depth of the optical fiber sensor area, and competition, most of NanoSonic's business shifted away from fibers and optics to chemistry, advanced materials, and applications.

Key steps in NanoSonic's growth came from efforts performed outside regular programs and in sharp contrast to planned project tasks. Several were related to the usual problems associated with ESA, namely, the length of time required to chemically form multiple monolayers and the spatial dimensions over which ESA coatings can effectively be applied. Both of these issues are related to the nano-to-macro problem that currently limits the application of nanotechnology to practical (human-sized) materials and structures. NanoSonic chemists and chemical engineers developed ways to overcome the perennial difficulties regarding the amount of time needed to form many-bilayer coatings as well as the maximum size of substrates that may be coated with good property and uniformity control. Multilayer ESA coatings up to a millimeter in thickness may now be formed in less than an eight-hour work shift, and substrates up to several meters square have been coated with good uniformity. Further, by first coating a substrate with a chemical release layer, then forming the coating, and finally etching away the release layer with acid, free-standing, self-assembled materials may be produced. NanoSonic's Metal Rubber is far from special polymer-based coatings on optical fibers, but it is an example of materials processing that is the current basis of much of NanoSonic's current business.

13.4 Metal Rubber Materials

Metal Rubber is a nanocomposite material whose properties can be varied over a wide range through varying composition and process parameters. The material is particularly interesting when configured as a stretchable metal-like conductor on an elastomeric substrate. Such conductors are required as mechanically flexible interconnections in polymer electronics, flexible circuits, electronic textiles, and a host of similar applications. Pure metal may exhibit very flexible behavior, exemplified in recent reports of a 100-nanometer-wide gold stripe evaporated onto a polymer substrate and maintaining finite electrical conductivity for substrate strains exceeding 20%. There are clearly practical issues with such systems, including long-term adhesion under cycling and the necessarily small size electrodes.

The resistivity of gold is in the region 10^{-6} Ω cm. NanoSonic has developed rubbers with resistivities down to 10^{-5} Ω.cm. These materials may be routinely strained to 100% while retaining their electrical conductivity, and they will recover their shape elastically.

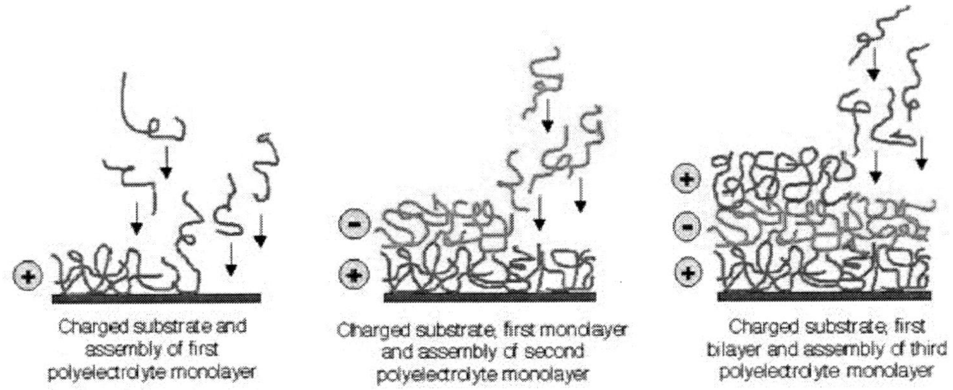

Figure 13.1 ESA process for the formation of multilayer thin films.

Typically, electrostatic self assembly (ESA) is used to form these materials (see Figure 13.1). A cleaned, carefully prepared substrate is immersed in a solution of the materials to be deposited. Depending on the charge on the surface, the positive or negative "end" of a polymer molecule adheres to the surface, leaving the opposite sign charge on the new surface. Through careful processing, ordered single layer stacks of molecules can be fabricated. Varying molecular composition within these stacks and/or adding dopants (e.g., gold nanoparticles) offers a huge variety of Metal Rubbers. Furthermore, the Metal Rubber can be either on the substrate material, which may be polymeric or fabricated as a thin sheet device by interspersing a release agent between the substrate, and the ESA deposited layer. Additionally, the Metal Rubber may be patterned through either surface treatment or etching processes to form contact layers.

Figure 13.2 shows a representative patterned flexible polymer substrate material. Here the dark lines, which are approximately 2 cm in width, have electrical resistivity of the order 10^{-4} Ω.cm. Figure 13.2 also shows a representative measurement of electrical conductivity for these particular deposited materials.

Metal Rubbers offer excellent performance as flexible electrodes with very high reliability over mechanical cycling. With slightly different formulations, Metal Rubber can be realized as a synthetic polymer actuator material with unparalleled strain range and strain recovery capacities. At NanoSonic, we expect that these Metal Rubber materials will find enormous diversity in applications within the near future.

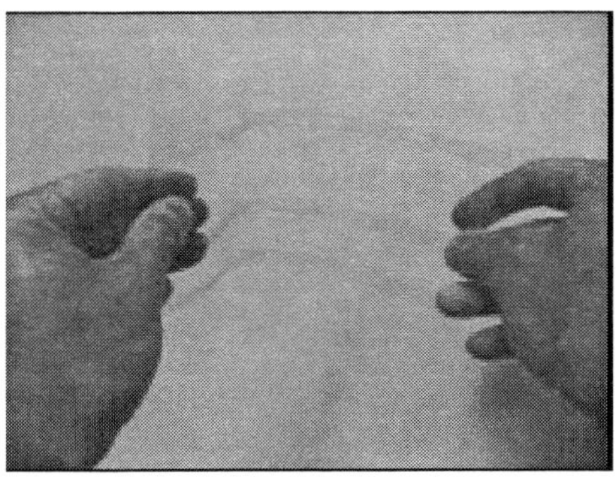

Figure 13.2 Patterned, electrically conductive lines on a flexible polymer substrate (left) and representative measurement of electrical conductivity (right).

13.5 Company Perspectives and Issues

NanoSonic is an example of a small company that has been spun off from a university research center by teaching faculty, research and administrative staff, and students. The company has grown gradually since it was formed by three individuals in 1998 and in mid-2007 had 65 full-time and part-time staff. It is positioned to build on its established intellectual property and the results of its research and development during the past decade. Some perspectives that may be of interest, either in their similarity to or differences from the experiences of other small companies involved in advanced science and engineering are as follows:

1. The spin-off process from the university was inefficient in that the university was largely uninvolved, and the company had to learn business practices on its own. The lack of an initial strategic partnership between the university and the company in the short term likely delayed company growth and in the long term likely reduced the potential positive impact of the company on the larger university community.
2. Space and infrastructure have been a constant issue. NanoSonic decided not to rent university-managed incubator space in Virginia Tech's Corporate Research Center for two reasons: university incubator space rental rates were approximately four times that of other space in the Blacksburg, Va., community, and incubator space did not include wet chemistry laboratory space needed for research work in chemistry and material science. Again, the lack of an effective, practical partnership between the company and the university community did not optimize growth. NanoSonic rented several spaces in Blacksburg and gradually built up its own lab facilities and equipment and office structure. Over time, and with the investment of substantial profit, NanoSonic's laboratory facilities are now better than any comparable labs on campus, other than very expensive equipment such as a scanning electron

microscope (SEM) and nuclear magnetic resonance (NMR). While lab facilities are state of the art, office furniture has been acquired on a parsimonious budget, saving much money and not affecting technical work in any way. Fortunately, Virginia Tech's surplus property organization was located a short distance from NanoSonic during most of its recent growth phase, and desks, chairs, and cubicle dividers were purchased usually for one or a few percent of their original cost. As a result, overhead costs were kept low.

3. Business assistance from university, local, and state offices as well as business suggestions from individuals and possible investment or venture capital groups was not found to be useful. Largely, NanoSonic found that individuals or groups offering to provide help did not understand, or try to understand, our technology and our group. Our best business assistance has been obtained from our attorney, accountants, and local bank.
4. Papers and conferences decreased significantly in value compared to their importance to university personnel, because unlike university researchers intent on showing their latest results to colleagues, company personnel are motivated to keep good ideas confidential. As a result, papers published tend to be summaries aimed at marketing, and conferences those where customers will be in attendance. Some effort is made to include young staff as paper authors to help document their efforts and grow their résumés. Rather than emphasizing paper publication, NanoSonic has been active in receiving awards for technical and business achievements, and these awards have been useful because they are visible to potential customers. NanoSonic was named the top small business in Virginia by Virginia Business magazine in 2006, received a NASA Nano 50 award for 2006 for one of the top nanotechnology products, received a R&D 100 Award in 2007 for its Metal Rubber textile materials and sensors, and has been nominated for the DARPATech 2007 Small Business Award.

13.6 Summary

NanoSonic is an example of a spin-off company from a university research center. It has transitioned some basic research initiated at Virginia Tech into commercial products and is well positioned to make significant future developments in molecular-level self-assembly and related materials science.

Bibliography

Arregui, F. J., Liu, Y., Lenahan, K., Holton, C., Matias, I., and Claus, R., "Optical fiber humidity sensor formed by the ISAM process," *Proc. 13th OFS* , Konju, Korea (April 1999).

Arregui, F.J., Liu, Y., Matias, I .R., Claus, R.O., *Sensors and Actuators B 3000*, (1999).

Arregui, F. J., Matias, I. R., Liu, Y., Lenahan, K. M., and Claus, R.O., "Optical fiber nanometer-scale Fabry-Perot interferometer formed by the ionic self-assembly monolayer process," *Optics Letters*, Vol. 24, No.9, pp. 596-598 (1999).

Decher, G., "Fuzzy nanoassemblies: toward layered polymeric multicomposites," *Science*, Vol. 277, 1232-1237 (August 1997).

Lacour, S., Wagner, S., Huang, Z., and Suo, Z., "Stretchable gold conductors on elastomeric substrates," *Appl. Phys. Lett.* 82, 2404 (2003).

Liu, Y., Rosidian, A., Lenahan, K., Wang, Y-X.,Zeng, T., and Claus, R., "Characterization of ESA nanocomposite thin films," *Smart Mater. Struct.* 8, p. 100-105 (1999).

14

The Story of Fiberonics

Jaspreet Singh
Founder, Fiberonics
Delhi, India

In this chapter, the creation and life of new high-tech companies in the context of a developing country is offered and useful conclusions for entrepreneurs are extracted.

14.1 Introduction

After obtaining a Masters of Technology degree in Optoelectronics and Optical Communication in 1989 from the nationally and internationally renowned Indian Institute of Technology (IIT) Delhi, my initial plans were to head to the U.S. for further studies (PhD/ MBA) after working in India for about one year, the U.S. being arguably the most favored destination for a majority of IIT undergraduates and graduates. At that time, a couple of major optical fiber companies were operating in India, both formed in collaboration with respective multinational corporations. Jobs were aplenty, as barely 10 students graduated every year with an Masters of Technology degree in the aforementioned area, and of course, IIT campus interviews ensured that everyone had job offers, in some cases even multiple job offers, well before the student actually graduated.

However, during early 1990, the Fiber Optics Laboratory at IIT Delhi had received a rather substantial UK Overseas Development Assistance equipment grant, which made me decide to stay at IIT and follow a different and challenging path: to utilize the equipment in developing indigenous techniques in fabricating fiber-optic components and also developing software packages to perform computer-aided experiments with some of the equipment to enable real-time characterization of optical fiber and components. Even the overseas manufacturers of the equipment to be used for computer-aided measurements had yet to develop the software, so obviously we did not have the option of sourcing it from anywhere!

Money by way of a salary was not at all attractive, and job security was certainly not there as the job was under a project appointment. However, the environment was very friendly, with encouragement and cooperation from peers, senior faculty, and junior staff being almost a regular feature. By the time I decided to move on from IIT Delhi after about three years, apart from achieving what I had set out to do, the biggest gain was the self-confidence obtained in working with different types of equipment, developing photonics software, and the satisfaction that these achievements were a result of my own efforts. Also, I had overcome typical problems such as summer power outages in New Delhi and problems related to importing small quantities of consumables for lab use. It is interesting to point out here that, apart from the rigors of paperwork required for local clearances, I clearly recall that even importing small quantities of Corning optical fiber (single moded at 633 nm) in the early 1990s required U.S. State Department approval, though I do not recall a single instance when the request was denied.

14.2 Initial False Starts

After leaving IIT, I opted not to join a PhD program in the electrical engineering department at Purdue University in 1992. I felt that this would set me back by at least three years in setting up a business, so I decided to set up a company of my own in the field of fiber optics. However, I was very unclear about the problems I would face, did not have any contacts in the business world, and more important, did not have any experience, something every well-wisher advised me was critical to doing business in this part of the world. The only savings I had were those from my salary during my three years at IIT Delhi, which were obviously not enough to set up a full infrastructure for starting and sustaining a company.

Business contacts in the fiber-optics field were also nonexistent, as the only major optical fiber companies in India at the time were those involved in manufacturing optical fibers and cables, and they had already established a network of suppliers within India and abroad. In addition, these companies were directly supplying their products to government-owned, public-sector companies (mainly state-owned landline service providers). Setting up an export-oriented unit was another option, but the lack of funds was a major constraint.

The only option was to wait and watch for an opening. To keep myself occupied, I decided to offer my services as a consultant to overseas companies wanting to set up operations in India and to Indian companies wanting to expand into fiber optics. It wasn't easy getting a consultancy assignment, as several interested overseas companies were already big players, entering India either on their own via their Indian subsidiaries or in collaboration with Indian companies.

For six months nothing happened, and much of my money was spent in correspondences with foreign companies. Eventually, the first "success" came—a U.S. company was looking for a consultant to help them enter the Indian Airports Telecom Sector as a prime subcontractor of AT&T, and I was contacted by their representative in India to assist him in creating a business strategy after identifying the various telecommunication requirements of the Indian Airports sector. The only problem was that I hadn't a clue as to what the major hardware and/or software requirements for the Airports Telecom sector were, or for that matter, what the Airport Telecom sector in India was aiming for! Anyway, I decided to take up the consultancy.

The first step was to establish contacts, and thus began my education in the field of business—getting to know people, a very important step, totally in contrast to my erstwhile job at IIT Delhi. Contacts were established within a month with the international and national divisions of the Airports Authority, and considerable self-study at the IIT Library (IIT alumni pay only $1 annually to use the wonderful library even today!) led to a reasonable understanding of the requirements and equipment for the airport telecom sector, so that at least I did not feel at sea conversing with the Airports Authority officials.

Then came the big damper. AT&T decided not to enter the airports telecom sector in India, as their focus was revenue generation in the near term, and the Indian airports sector only offered projects likely to generate revenue in the mid- to long term. The first consultancy ended without much success, but there were some positives—a reasonably good learning experience on how to create contacts quickly in an altogether different field, and the important fact that bringing in new technology would take time in a country like India, regardless of the international brand name involved. The money earned was next to negligible, as everything had ended very quickly.

I was determined to prove to myself that I could sustain myself through consultancy assignments, and the second one was offered by a neighborhood company that wanted me to help them streamline their operations in a $1-million construction project in New Delhi. The company was facing big losses by way of construction site pilferage, and the pace of construction was far behind schedule. Again, this was a totally alien environment for me, but I decided to take up the assignment in my enthusiasm for learning something new.

Some of the mistakes I made were that I did not

- Get a contract prepared by legal experts for binding the company to pay me at fixed intervals, based on the quantity of work I had done.
- Negotiate a productivity-based remuneration package, as opposed to simply accepting a flat consultancy fee at the end of the consultancy period.
- Insist that the company insure me against any mishap at the construction site or while traveling. I did not even have an accident policy then, which was very foolhardy considering that accidents are very likely at construction sites, or even on roads, given the wonderful traffic conditions we have in New Delhi.

Oblivious of the mistakes as I commenced the consultancy assignment, I realized that construction project management was quite a handful. I had to deal with a hierarchy of employees having different educational backgrounds, ranging from engineers and architects to the labor force, which probably did not even have a proper high school education. Additionally, at this particular site, the work hours seemed to vary every day. Obviously, a major revamp was required to get the project on track.

The first issue was to get myself recognized by the work-force at the site—engineers, architects, and labor force—as someone representing the management instead of a mere consultant employed by management. After achieving this goal, which took about three weeks with the help of the company president-cum-CEO, the rest was easy; all I had to do was weed out the corrupt and inefficient, reward the efficient workers/subcontractors, and ensure that pilferage was reduced to a minimum, which I accomplished by making all the subcontractors accountable for whatever material was issued to them and requiring company officials to perform verification in my presence every evening.

The major experience I gained was the realization that the fundamental contributors to a successful business strategy are common sense and employee goodwill.

The project, which was running six months behind schedule before I came on board, was completed on time. After the usual round of congratulations and expressions of gratitude by the management, I was fairly upset that the company refused to pay anything for the additional tasks I had helped them with not related to the consultancy assignment, such as negotiating prices with some key suppliers, conducting technical appraisal of certain items, settling disputes of the company with a couple of suppliers, etc.

The best I could do was to voice my disapproval because I realized it would be futile to take a big company to court (Indian courts are well known for taking ages to dispose of lawsuits!). The company offered me some freebies like a partnership in a new business they wanted to set up with my assistance in the area of fiber-optic lighting for buildings, including allowing me to operate from their premises without charging rent, but the adage "once bitten, twice shy" seemed to hold true for me.

14.3 Setting up Fiberonics

Realizing that consultancy assignments in India did not pay well, since the concept of paying for the consultant's time seemed to be lacking, I decided to take the plunge and set up my own company in 1997, regardless of the problems envisaged. I named it Fiberonics, as I was sure that my focus was to work in photonics, mainly in the area of education because fiber optics was slowly being introduced in the curriculum of several universities. However, the first task was to raise money.

To do this, I decided to target the middle-class population (comprising mainly the salaried and the reasonably well-off business classes) in the city where I resided—New Delhi. The city faced (and still faces) a perpetual water problem during summers, and with much of the population living in four-story apartments, almost everybody relied on privately installed booster pumps to pump water from the mains to the respective overhead storage tanks. Many people simply relied on hearing the sound of overflowing water from the overhead tank as a signal to switch off the booster pump! Obviously, a better solution was to install a controller or alarm system to serve as an alert that the tanks had been filled.

Hence, I decided to go in for manufacturing an advanced electrical/electronic water-level indicator and booster-pump controller. Additionally, I felt that such a device would help curb the colossal waste of water, especially during the summer water shortages, which was also consistent with my thinking that engineers should contribute toward saving the environment.

I roped in a cousin who was working with a company manufacturing furnaces to help me with the manufacture of the water-level indicator and booster-pump controller. The business arrangement was that he would be paid 50% of the profit, though I realized later that it should have been closer to 25%, considering the time inputs, as he did not want to give up his job.

The first step was to conduct a market survey, and we started from the nearest hardware shop. We found that two brands were already in the market, and one had very sleek, attractive packaging. The costs: the sleek one was about $40 and the other just under $20, exclusive of the external wiring costs. We also noticed that the relays used in both were locally made and the wires were of inferior quality, which meant there was no quality control. In addition, we inquired about the quantity of sale, and were informed that the respective products did not sell too well.

I decided to frame my own product development and sales strategy, and I planned as follows:

1. Not to involve any hardware stores for sales, as the only thing that matters to a shopkeeper is the sales commission, not the product's technical capabilities. Also, even technically competent shopkeepers prefer not to spend time discussing the technical superiority of one product over another unless the store just deals in one type of product. Therefore, the strategy was to target the consumers directly, specifying the major advantages in saving water and electricity costs. Indians by and large are immune to the concept of "save the environment," so the decision was to target the monetary benefits to the consumers.
2. Prepare a good product flyer, which we could hand over or mail to prospects.
3. Use the best branded components available in the market, so that we could
    a. Maintain proper quality without worrying about calibrating components each time,
    b. Minimize after-sales servicing, and
    c. Ensure continuous component supply.
4. Decide to keep profit margins towards the lower side initially because of the existing competition.

5. Use the money generated to improve the design and packaging periodically, as opposed to splurging it on dinners or vacations!
6. Fabricate the product in my garage at home as opposed to setting up a unit in a commercial area where we would have to pay rent, which would overload the already-meager resources.

It proved to be hard work to locate the component suppliers, as we had to visit several shops in the biggest electrical and electronics markets in New Delhi, located near the famous Red Fort. Unfortunately, all these shopkeepers dealt in wholesale, so for a few pieces, the quotes were close to the list prices and bargaining did not seem to work.

As we did not have orders in hand but were sure we would get buyers easily, we initially manufactured five units. For packaging, we used off-the-shelf metallic boxes, usually used as junction boxes. Though the packaging was not attractive, our efforts bore fruit, and we immediately got two orders simply by directly promoting the products to customers. Both customers wanted custom-designed options to operate multiple booster pumps, and we were easily able to deliver what the clients desired. Considering that we offered custom-designed versions not offered by any other vendors, we demanded our price—and got it, too!

We managed to sell all the systems by pricing the product midway between the aforementioned commercially available systems, and orders (though slow) started coming in, enough to sustain the business. However, after the initial successes, I found that keeping low profit margins was taking its toll on my partner's interest. He felt we should charge substantially more. He was right in one respect; installing the systems was labor intensive, and we were not charging for the time for each installation, including time and money spent in traveling to the clients' premises. I realized that we would not get any orders if we charged clients for the installation time, so we had no option but to change the design. The new design did reduce the component costs, giving us some extra margin, but not much.

Eventually, my partner quit. He got a job offer from a multinational company and decided he did not need the extra money that came from selling our system. I carried on independently for a while, executing orders on my own with the help of one technician. Eventually, another company entered the fray and started selling a cheap system at $10—barely the price of the contactor used in our system! Though I realized we could not compete with the cost, I decided to continue selling the system at the price fixed earlier for custom orders, but sustaining the company was turning out to be very difficult and I desperately needed another source of income to offset employee salaries and office costs.

14.4 Creating an Additional Source of Income

Just about the time Fiberonics was established, I met Prof. Brian Culshaw of the EEE Department at the University of Strathclyde during one of his visits to India. I had known Brian since my student days, as I used to attend his seminars during his visits to IIT Delhi. I must also mention that Brian's contribution over the years in helping IIT Delhi in the field of photonics education and R&D is acknowledged even today by one and all at IIT Delhi!

During this particular visit, Brian was planning to involve the Fiber Optics Group at IIT Delhi in a UK-India science and technology research project. Brian's proposal was to initiate interest in optical fiber sensors for the Indian civil infrastructure, and the project guidelines required a tie-up with industry both in the UK and India during the entire tenure of the project. I jumped at the opportunity and offered to serve as consultant for the industry link, as this was a good opportunity to enter a new field. There was no consultancy fee or remuneration to any of the industry partners, but the project did allow the Indian industry partner to visit the UK.

The project's principal investigator from IIT Delhi, Prof. Bishnu Pal, suggested to Brian that there should be some additional source of income for the industry partner in India, and Brian offered me the opportunity to represent his company, OptoSci Ltd., in India to promote photonics educator kits to Indian universities as a means to generate some income.

This meshed perfectly with the plans I had for Fiberonics, and I immediately agreed. The task did not require any training in the UK, and it was easy to promote educator kits by telephone, post, and email in India, considering we already had an operational office. Since I had a post-grad degree in optoelectronics and optical communications from IIT Delhi and had worked there, it was easy to tap existing contacts and build new ones for suggesting relevant equipment for coursework.

However, it took us two years to get our first order (during which I had to sustain myself on the booster-pump controller business). The main problems in selling fiber-optic kits to universities were

1. Low levels of available funding, mainly due to heavily subsidized education (Interestingly, IIT charges less annually for professional education than privately owned primary schools charge for primary education!).
2. Photonics education was still in its infancy and faculty was generally not available at universities across the country.
3. A tender process was required to be followed for each purchase above \$4,000, and every OptoSci kit was priced above that mark. Normally, tender processes can take over a year to complete in India, depending on several factors.

Once we started getting orders, there was no looking back. The biggest highs have been getting repeat orders from our clients, some almost on a yearly basis, which obviously means that we have been successful in generating customer confidence both for the products and the support available in India. From one sale a year initially, we have been able to generate multiple sales with some clients ordering multiple kits.

In several instances, I have had to help university faculty prepare funding proposals from government agencies. True, I do not charge them for the time I must spend in drafting and finalizing the proposals, but it is very important to convey to the universities that they can get a lot of value from us by way of sound technical inputs in India, both with regard to the proposal and during project implementation. This strategy gives us a clear edge over our competitors in India, who at best come up with sales talk and usually cannot provide technical inputs on their own. In addition, I have never refused to deliver seminars on popular topics in fiber optics, whenever invited by universities in India, as it improves our profile and gives us additional publicity.

The other thing I must highlight here is the excellent teamwork we have established with OptoSci. It's great when the OEM can customize products depending on the needs of an institution here in India, based on our input, apart from selling the standard product range. Dr. Doug Walsh, OptoSci's general manager and director, has been a delight to work with over the years, and apart from the normal support he has been instrumental in offsetting several out-of-pocket expenses such as internal travel, hotel expenses, advertising costs, etc. This has given us the incentive to go that extra mile in generating business for OptoSci.

In 2004, I also developed a software package for waveguide analysis in India, which is now sold internationally by OptoSci because it supplements one of the educator kits and is under the joint copyright of OptoSci and Fiberonics. The software took about six months to write and was tested by OptoSci's product development manager, Dr. Iain Mauchline; one of Iain's most interesting observations was that my software failed to address the French decimal system! Eventually, I found a solution for that, and the software now operates in conjunction with the regional settings of the Windows OS. The fact that I sell the software via OptoSci has helped me reduce costs that I would otherwise have incurred in advertising overseas or setting up a Web site.

Our next target is to promote OptoSci's methane gas sensing system in India for mines and landfill applications.

Fiberonics has also recently partnered with VPIsystems as their business partner in India. Undoubtedly, VPIsystems' products are far superior to similar products the competition offers, but we haven't been able to generate orders to date because the products are priced higher than the competition, and the Indian faculty feel that the software may be too advanced for the university segment here. Selling software is not easy, as we have learned the hard way, but we are working on a couple of new strategies now to realize sales.

All in all, being in business has been a roller-coaster ride, but the satisfaction of owning your own company is far greater than working as an employee. True, there is no paycheck at the end of every month, and one has to struggle almost on a daily basis, but one has to be up to the challenge if one wants to set up and own a company.

14.5 Key Points for a New High-Tech Company

In the end, without aiming to sound didactic, I would like to include a checklist of important points for aspiring high-tech entrepreneurs, based on my experiences over the past eight years or so (though I am still learning!).

1. If you plan to start a manufacturing unit, focus on one or two specific products, after carrying out a preliminary market survey/competitive analysis, and finalize a steady source of raw material from a competent source. If you plan on representing a company, focus on one that is not very big but has a good product range in addition to having technically sound professional management.
2. Plan on having more than one source of income, if possible, during the formative years of your business. It reduces stress, especially if you have a family to support.
3. Build useful contacts constantly in the business world and in relevant government organizations.
4. Try to work out the additional expenses (e.g., those for sales meetings, installation, etc). These are definite costs because they involve time.
5. Try not to out-source sales until your sales rise to large levels.
6. Technical competence is necessary, so do all you can to learn.
7. Work out all the legalities and consult a good lawyer to draft business/partnership agreements.
8. Employee loyalty is important, so make sure your team is always satisfied (most difficult!).
9. Never pay kickbacks to get orders. You may get some quick successes initially, but eventually your reputation takes a beating. In our case, at best, we may have lost one order over the past eight years by not paying kickbacks. The quality of the products—with a sensible marketing strategy and technical competence, rather than under-the-table means—must define sales success.
10. Work hard and never give up!

15

Founding a Fiber-Optic Component and Sensor Business

Ingolf Baumann
Founder, Advanced Optics Solutions GmbH
Dresden, Germany

When an engineer is going to start his own business, only a few people are able to give some competent advice based on their own experiences; most entrepreneurs might face this actuality. Since the primary business idea is related to the high-tech business sector, it may even become a real challenge to gather all necessary information that enables the founder to realize his ideas. However, the founder has to overcome this first barrier, being much more on his own compared to a business creation in a more traditional field.

The basic reasons for this are the extremely high customization effort for most innovations, serving the market's demands, and a lack of established distribution channels for the product. Thus, in this state it is quite difficult to prove that the company is making progress and that the business plan is a realistic scenario.

As a matter of fact, no other person is as deeply involved in the young company's technology and/or services as the founder—only this competence might be able to dispel the investors' doubts. The following lines may give an overview about my way of creating a successful business.

15.1 Founding

After completing my doctoral studies at the photonics group of the Dresden University of Technology, I founded the company Advanced Optics Solutions (AOS GmbH) in 1998. Because of this scientific background, the main focus of the young company was to be a special optical-fiber component, called a fiber Bragg grating (FBG), for the booming telecommunication industry. FBGs are optical filters that are inserted directly into a glass fiber, transmitting and reflecting optical signals within a narrow spectral band. These components were considered as possible key components in telecommunication systems, which benefited from incredibly fast development in the late 1990s.

Furthermore, FBGs can be used as fiber-optical sensors for temperature and/or strain. That feature made the component even more attractive as the key point of a business idea because of its possible use in a completely independent second market. Up to today, the gratings are still a main product and an essential part of the company's activities.

The key component for telecommunications with FBGs was the optical add-drop-multiplexer, using the gratings for wavelength separation of optical communication channels [i.e., in Dense Wavelength Division Multiplex (DWDM) systems]. This component was described theoretically, and the first experiments were very promising.

This idea was accompanied by a couple of other applications, such as laser-diode stabilization, fiber-laser applications, ASE suppression, and a few other ones.

The sensor market was the second field we entered. Fiber Bragg gratings were known to change their optical properties under mechanical stress or temperature change, shifting their central wavelengths in accordance to the grade of impact. On the other hand, they are immune to electromagnetic fields, chemically inert, and have an intrinsic long-term stability. So AOS developed the first fiber-optic sensors based on fiber Bragg gratings for applications like civil engineering, structural health monitoring, power plant surveying, and geo-techniques. Following the market's demand, a portfolio of monitoring units has also been created, providing support for most FBG-based sensor systems.

This way, AOS was going to enter a couple of small niches in two huge business fields, equipped only with an excellent scientific background in fiber and a couple of strategies to serve the two different markets.

15.2 The Business Environment

The method for financing a start-up depends on the objectives of the business idea. The financial resources of the founder are usually limited, especially in the high-tech sector, which is known for its immense need for investment. Private investors and their venture capital are needed in most cases when a company is established to speed up the founding process and prevent the young company from losing too much time. A delayed founding process may rapidly decrease the competitive edge in terms of the company's advances in research and technology compared with its potential competitors, resulting in the inability to generate turnover before these competitors will. But for private money, the fast generation of a high turnover and the rapid return on the investment is nearly the sole objective of participating in a founding process.

This contains a lot of well-known risks that can easily lead to a sudden collapse of the young company. Therefore, for me it seemed to be more appropriate to co-finance the start-up with a public research project. Compared to private investors, public money that supports private business activities aims for a slightly different kind of revenue: eventually, both want to earn money. However, a public investor is also interested in creating jobs and a stable tax income, so the long-range forecast is often more important than the fast return of investment.

Bureaucracy is necessary. During the founding of a company, this is sometimes forgotten, but you will be reminded. When I wanted to register the company with the legal form "GmbH" (comparable to the Ltd. designation), the attorney wanted to know where the company kept its principal office. He explained that a tenancy agreement for the office and the production space would be necessary to proceed with the registration.

So I went to a new business center that had some production and office facilities to rent and asked for the rental agreement. I was told that I would need a valid registration number of an already registered company. Considering that such precarious situations dealing with legal actions require sensitive and innovative action, we arranged a meeting between the registration office and the rental office. Eventually, together we found a solution.

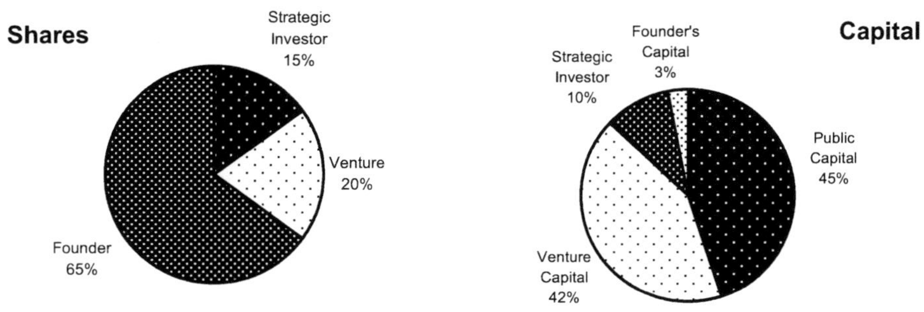

Figure 15.1 Shareholders and first financing round for AOS.

AOS got the chance to rent a working area in a new, big office building as the first company. The intention was obvious: in new business buildings, it is often difficult to find tenants for the first 20 percent of space because nobody wants to work in an empty house. AOS signed the contract and, moreover, the company was granted a really low rental fee for the first three years. One moment later, the legal registration was finalized with two signatures, one by a very satisfied attorney and one by me, while the facility officer filled the registration number in the contract.

Another emphasis of the founding process was to find a suitable staff for the start-up company. I found many books and a lot of different philosophies about the right selection of personnel and human resources management. A lot of them were only applicable to larger organizations. In a small company, the CEO usually has a good knowledge of each employee's current activity and the current state of his work.

At the beginning of the founding period, we started with two well-qualified engineers I had already worked with at the university over many years. But soon the company's growth enforced us to increase AOS' personnel. One important criterion for employee selection (besides the qualification in special fields) was the candidate's ability to work within a team.

Although a fair competition between employees can be absolutely beneficial for a company, contrary opinions or attitudes among colleagues must never lead to intrigues and/or mobbing activities. In every company, the management is obliged to point out clearly that there is no tolerance for scheming against each other. The manager's ability to moderate and mediate in conflicts may help a lot in keeping productivity at the highest possible level, even in difficult economic periods. During the telecommunication crisis in 2002/2003, the business was getting more awkward for AOS and likewise other companies in this industry. The continuous growth stopped abruptly, and the turnover in the telecom sector shrank to less than half compared with the year before. In this situation, my decision to focus on soft skills was actually honored. The excellent team spirit and strong cooperation between the management and staff was of great help for AOS.

A well-adjusted amount of transparency ensured the high motivation level that was necessary to be with a young company that found itself in a commercial environment drawn by daily collapses and bankruptcies. I personally believe that this was an effective way to help the whole staff face the upcoming new challenges with confidence. Up to today, we still practice an open information policy, containing the communication of the business objectives, the current market situation, and the management's major strategic decisions.

For a young start-up company, management must provide a strictly defined set of competencies and responsibilities for each employee. By doing this properly, the founder will very seldom feel impelled to correct the employee's decisions. With AOS, we have practiced this way successfully over the years.

When preparing the business plan, most people forget to plan for attorney costs. Sooner or later, they will need the help of an attorney. Immediately after the registration, AOS had to obtain a license to commercialize its products. The draft of the contract was quite a comprehensive work. However, it was much more complicated to understand all parts of the content, even after the translation. Therefore, for the subsequent communication and the final contract negotiation, there was no alternative to commissioning a law firm's professional service. Due to the attorney's specialization in intellectual property affairs, it took "only" six months before the firm got the license. After this process, we were sure that it would never have been possible for us to go this way alone. From the commercial point of view, the planning for the attorney costs in the business plan has been a wise decision.

15.3 Experiences

Right from the beginning, we had to identify appropriate marketing channels for AOS. A new company with innovative products is first and foremost obliged to establish its name in the world of business, accompanied by a winning reputation as fast as possible. Naturally, any start-up company has more or less limited capacities to check every possible marketing strategy that is shown in publications. For AOS, we selected participation in international conferences with affiliated trade shows as our first sales-related activity. On the one hand, we had the opportunity to get in touch with our customers and promote our first products along with the opportunity to take a closer look at our competitors. On the other hand, the presentation of the latest research results during the conference could be used effectively to verify our strategies and adjust our product ideas.

The first customer inquiries we received this way were widely spread among a lot of application fields, despite the fact that most of our marketing efforts were targeting standard industrial applications within our primary business field—the telecommunication market. Some customers wanted to get a very special solution the next day; others asked for our participation in large research projects that were going to start over the next few years.

Sometimes it was obvious that a customer merely requested a third offer, with the sole intention of gathering the required number of quotations for their procurement offices, while he had already made his decision to purchase from a competitor. Bidding for such tenders would definitely result in a loss of money and time.

However, the identification of such activities was a long learning process. First, it was quite overwhelming, and it soon turned out to be very difficult to measure every inquiry's importance reliably. One question had to be answered repeatedly: Would the requested solution lead to a real product with at least a possibility of being accepted by a real market, or was it a one-sample-order? With time, we have learned to identify customers who were interested in collaboration and cooperation, and those who were only looking for a second source or a one-time job.

Beside our attempts to serve our new customers as optimally as possible, we were continuously communicating with shareholders, the house bank, the licenser, the rental company, etc.

During these first months, the daily work absorbed me completely so that I was afraid of getting lost from time to time. Due to this large amount of work, in the first year it was nearly impossible to verify the company's strategy. We hardly found time to make at least short monthly evaluations of economic efficiency. Often there were about half a dozen topics to investigate in order to set things higher or lower in priority, and any resulting decision could influence the further development of AOS in a good or bad way. One of the most significant issues is how to weight strategic decisions, focusing on long-term market development or short-term profit. During the crisis in 2001, we learned that a good balance between these two antagonists is mandatory.

In the beginning, there is absolutely no guarantee that your business idea will be profitable at a certain date that also looks reasonable to your investors. In the late 1990s, the telecommunication market especially got a remarkable boost, mainly through the break-through of some important key technologies such as data transmission over an optical fiber. We were aware that we would probably need about one year to fine-tune our technology to produce fiber Bragg gratings, and one additional year to produce the first FBG-based component, the add-drop-multiplexer.

But what of predictions about where the market is going in the next two years? The next year's weather forecast might have been more reliable. People with a technical background often have mixed feelings about making predictions without any confidence, while people with a commercial background tend to mix up confidence with assumptions. Anyway, we had to observe the market forecasts carefully, continuously looking for alternative solutions and applications that probably could be served by our core technology. Doing this permanently was sometimes annoying but eventually we managed it successfully; after about one year we established a small portfolio of competitive products that generated a noticeable turnover.

On top of the boom in 2000, it was quite easy to raise funds for a company that was related to the telecommunication field. Thus, it was easy to get money to expand our capacity in order to produce fiber components in high volumes. Bankers gave positive assessments immediately when they heard that the company dealt with telecommunication products because of recent years' statistics and their proven inability to recognize that the market had already been overheated. Ignorance of the old businessman's rule that if you could produce several hundred products every hour you would also have to sell several hundred per hour led to that massive demise of young companies when the bubble burst one and a half years later.

Just in time, automatic production stages for fiber Bragg gratings became available commercially. This acquisition meant a huge investment for our young company. On the other hand, it was now possible to noticeably increase the production lead of standard gratings. However, soon we recognized that the all-in-one production lines were limited in terms of flexibility. But flexibility was one of the major market demands, and with AOS we were proud to count it among our core competencies, so we were able to produce standard gratings with a high repeatability while merely covering about 20% of our average order income. For the remaining 80%, we still had to use our semi-automatic production stages. In accordance with this experience, we refrained from further investments in full automatic production lines. This decision was not approved by a lot of "experts" promoting their contemporary New Economy business models, but eventually we were right: two years later, we replaced a lot of our equipment with only a fraction of the original investment, simply by participating in sales auctions of former competitors who had broken down through excessive growth during the booming years.

When the market started to plunge in 2001, we immediately stopped development of the add-drop-multiplexer and also cut most of our other telecom-related activities. At the same time, we doubled our efforts in the optical sensor branch, which shaped up as more or less immune to the turbulence in the communication sector. First, September 11 and its impacts strongly affected the global economic condition. We were not able to escape completely from this negative development, and we also lost a lot of money. However, as a company we have remained largely unaffected.

Our customers, like those of other high-tech businesses, can be characterized by consultancy effort, ordering volume, and sustained yield, and the relevance of their orders will be scaled up or down accordingly. But that evaluation should be done with great care because the above qualities might change with time. In order to illustrate how difficult it is to find customers as well as how easy it is to lose them, I want to give one example.

One of our customers ordered a strain sensor and rented a monitoring instrument for an application he didn't tell us much about. It was merely a small order. We didn't get any feedback about his achievements and/or our product's performance so the contact ceased.

Eight months later, we discussed strain measurements in civil structures with another customer. His intended project was of considerable size and turned out to be a real challenge in terms of feasibility. We were surprised that from the beginning he was committed to the optical technology and told us that he already obtained good results with it. Later, it was revealed that he had cooperated before with the customer we lost contact with. This example shows that even a very small customer might act as a multiplier for volume, which is also a good chance for entrepreneurs with limited marketing capacity to gain profitable projects.

15.4 Conclusions

During and after the founding process, management is very busy with daily work. However, business has to be continuously adapted to the market's movements.

- Do not hesitate to stop projects that are later proven not to fit into the long-term strategy. You may lose money, but otherwise you will lose even more money.
- When you're going to change anything inside the company, think carefully about what you actually want to change and in what order. The number of attempts is limited by money and time.
- A lot of money, time, and patience are necessary for customizing products and solutions and for successfully entering a diversified market.
- Understand that the founding of your business is a process that spans a couple of years.

Acknowledgements

The figures and title in this paper are copyrighted by Ingolf Baumann and or AOS GmbH and used with permission.

16

How to Start a Small High-Tech Business in Troutdale, Oregon

Eric Udd
Founder and President of Blue Road Research
Oregon, United States

Blue Road Research grew from a one-person company in 1993 to 14 employees by the summer of 2002. This paper describes the core philosophy of the company and outlines its growth during this period.

16.1 Starting a Small High-Tech Business in Troutdale, Oregon

The motivation for creating Blue Road Research arose primarily from my wife's parents having health problems and her wanting to move back to the Pacific Northwest. Further complicating things were her parents wanting to sell the home she grew up in, which was in an unincorporated area of Oregon with a Troutdale address. Looking at the area around Troutdale, it became readily apparent that the only way to continue working on fiber optic sensors and live in Troutdale would be to create my own company.

Since my wife wanted to live in the house she grew up in, that pretty much defined the site. By selling our house in Southern California, where housing was relatively expensive, I determined that we could build a building to house the business as well as remodel and fix up the Troutdale house, which was in relatively poor condition. Fortunately, the site had 2.9 acres of land, so room to expand was not an issue.

The next step was to build a business model that would allow an adequate income to live on and eventually result in a self-sustaining company. There was also the issue of choosing a name for the company and putting the legal papers in place. These efforts took place concurrently over several months prior to my leaving McDonnell Douglas in August 1993. Perhaps more than any other business-related question I have had over the past eight years is how the name Blue Road Research arose. The process started with my outlining perhaps a dozen potential names. "Blue" was chosen because it is my favorite color, "Road" has to do with *The Wizard of OZ* (my favorite series of books when very young), and the "yellow brick road" that leads off to ???. Basically, I felt I was setting out on a road that I hoped would lead to something great, but it was definitely filled with ??? Finally, "Research" was added because that is what I wanted to continue to do, although the mission I defined for the company was to move fiber-optic sensor technology from concept to the field. My intent from the very beginning was to facilitate the transfer of fiber-optic sensors to fielded commercial applications.

To accomplish this mission, the model shown in Figure 16.1 was developed. It also was designed so that there would be sufficient income to allow us to achieve a stable financial condition before our savings ran out. From the customers' point of view, this model is intended to serve the end-user first by allowing a general introduction to the field of fiber-optic sensors. The intent here is to create customers that are as knowledgeable as possible and to introduce a new technology. By offering courses at reasonable rates, an easy entry point for potential customers was created.

The very first courses offered by Blue Road Research also offered laboratories to show customers how to build and use fiber sensors and associated technology. Blue Road Research has offered these courses continuously since its founding in 1993 and continues to improve on the courses. To complement the course, the Pacific Northwest Fiber Optic Sensor Workshop was created to promote the technology in the Pacific Northwest.

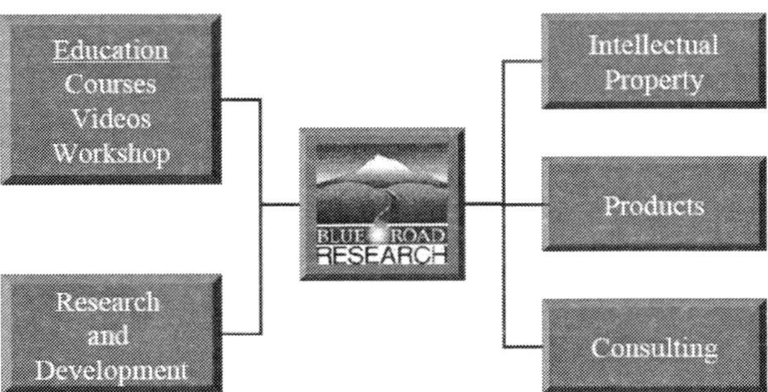

Figure 16.1 The Blue Road Research business model: It is designed to accomplish the goal of moving fiber-optic sensor technology out of the laboratory and into commercial products and applications. In the very early years, education and consulting dominated. R&D activities started to dominate after three years.

The second part of the business model involved offering consulting services to customers who needed more than a general introduction to the field. During the very early years of Blue Road Research, the majority of the revenue generated by the company derived from these two activities. By the end of the second year, a significant portion of Blue Road Research revenue was generated by research and development contracts from commercial and government sources. Initially, commercial contracts were dominant; but by the end of 1996, Blue Road Research won its first Phase II SBIR, and since that time government contracts have played the major role in funding Blue Road Research growth.

Since the company's inception, it was recognized that, in order to have a viable product base, intellectual property would be essential. Through negotiating with McDonnell Douglas, 18 patents were licensed with sublicensing rights obtained on patents generated by Eric Udd while at McDonnell Douglas. Seven other patents associated with fiber-optic gyros and secure fiber-optic communication were not licensed because other companies already had or were in the process of negotiating licenses. In addition, patent filings were made on a continuous basis to strengthen and expand the overall patent base.

After surveying all fiber-optic sensor technology and reviewing the patent base, it was determined that fiber sensors based on fiber gratings and the Sagnac interferometer would be the initial principal target product areas for Blue Road Research. While there have been time periods when research work on the Sagnac interferometer was a major part of Blue Road Research funding, products based on fiber grating sensors and their application gradually become the dominant focus of the company. This in turn resulted in the generation of products such as dual- and three-axis fiber grating strain sensors, educational and industrial kits, and fiber grating demodulators for high speed and sensitivity.

By the 1996/1997 time frame, Blue Road Research approached a break-even condition, and by 1998 product sales were beginning to become a significant segment of Blue Road Research business. By mid-1997, the company reached the break-even point and ceased to require additional investment to sustain operations. The performance of the company continued to improve in 1999 and 2000, eventually resulting in the acquisition of Blue Road Research by Standard MEMS in January 2000.

In 1999, and more aggressively in 2000, Blue Road Research was approached by larger companies interested in acquiring the company. At this point, the question became determining what the best next step would be. Standard MEMS seemed to offer the best prospects for growing the company to the next level in terms of technology and expertise in raising capital. As a result, effective January 2000, Blue Road Research became a wholly owned subsidiary of Standard MEMS.

16.2 2002

Blue Road Research over time was focused on six principal market areas: aerospace, civil structures, oil and gas, naval applications, composite materials, and environmental sensing. There were also some activities in the communication and biomedical field. Fiber grating products that were offered included single-, dual- and three-axis fiber grating strain sensors; fiber grating strain sensors packaged for civil structure applications; high-speed demodulation systems for fiber gratings operating with speeds of 10 kHz to 2 MHz; and custom and standard kits to support testing with fiber gratings. Major new products that were field tested included a high-performance fiber grating pressure sensor being developed for NAVSEA in collaboration with Schlumberger, and fiber grating moisture sensors field tested at Sandia Albuquerque.

Blue Road Research cooperated with a number of companies and organizations on projects that have system-level potential on aircraft and spacecraft, naval applications, civil structures, and environmental sensing. In the civil structure, Blue Road Research installed fiber grating sensors into the Horsetail Falls Bridge in 1998, the Sylvan Bridge in 1999, and the I-84 freeway in 2001. The Horsetail Falls Bridge was the subject of extensive static testing to verify the performance of a composite overlay method to strengthen the bridge using 28 embedded, long-gauge-length fiber grating strain sensors in the concrete beams and the composite overlay. This bridge was also used to demonstrate the ability to measure the speed and weight of traffic on the bridge as well as the presence of joggers. The Sylvan Bridge, which was scheduled for replacement in three to five years, served as a test bed for fiber grating strain sensors. A National Science Foundation (NSF) Phase II proposal was pending that would allow Blue Road Research and the University of California at San Diego (UCSD) to develop a system to monitor seismic damage to bridges and buildings in real time. This effort was to be supported by the Oregon Department of Transportation and Cal Trans. Blue Road Research, with support from the Oregon Department of Transportation, installed fiber grating sensors into the I-84 freeway. These sensors have been used since August 2001 for vehicle classification studies. Figure 16.2 shows an overview of some of these efforts.

Figure 16.2 Examples of civil structure system applications: (a) Horsetail Falls Bridge, (b) bridge bearing using multi-axis fiber grating sensors, (c) static and dynamic bridge load tests, (d) fiber communication cable installation to fiber grating sensors on I-84 freeway, (e) cutting grooves into I-84 freeway, and (f) installation of the fiber grating sensors for vehicle classification.

Several field tests were scheduled for the fourth quarter of 2001 and the first quarter of 2002 for the measurement of multi-axis strain in adhesive joints, moisture sensing on soil-capped hazardous waste areas, an underwater field trial of the Blue Road Research fiber grating pressure sensor, and additional civil structure testing on the Sylvan Bridge and I-84 freeway. Other tests were scheduled with government and commercial organizations on fiber grating sensor projects on which Blue Road Research was cooperating.

16.3 Ownership and Management Changes in April 2003

Standard MEMS, which acquired Blue Road Research in January 2000, went bankrupt. Blue Road Research, as a wholly owned subsidiary, was auctioned as an asset in April 2003. The successful bidders were Nick Ortly, the former president of Standard MEMS, and Peter Gatti, the former controller of Standard MEMS. The philosophy and management of the company changed radically from one of investment and growth to that of extracting funds as rapidly as possible. The company began to collapse in mid- to late 2005, and in January 2006 Eric Udd left the company. He was followed over the next few months by almost all former Blue Road Research employees.

16.4 Summary

From 1993 to 2002, Blue Road Research grew by developing and successfully applying its business model of offering customers end-to-end support in moving fiber-optic sensor technology to products and the field. It grew its product line and customer base, allowing it to fulfill its mission of converting the dream of moving fiber-optic sensor technology to real-world systems. In 2003, new ownership and management took over, focused on maximum returns, and the company collapsed in 2005/2006.

Acknowledgements

The figures and title in this paper are copyrighted by Eric Udd and/or Blue Road Research and used with permission.

17

SMARTEC: Bringing Fiber-Optic Sensors into Concrete Applications

Daniele Inaudi and Nicoletta Casanova
Founders of SMARTEC
Manno, Switzerland

Ten years after its launch, it is interesting to look back at the history of SMARTEC and see how a research project at the Swiss Federal Institute of Technology in Lausanne was turned into a successful company considered a leader in the domain of structural health monitoring, now part of a multinational group. We believe that the lessons learned in this process could be of interest for other researchers who have developed a new technology and feel the same entrepreneurial enthusiasm. We will show how the success of SMARTEC resulted from the combination of technology, partners, service, and sales strategy.

17.1 SMARTEC's Pre-History

The pre-history of SMARTEC SA started in 1992, when Prof. Léopold Pflug launched a pilot project of fiber-optic sensors at the Stress Analysis Laboratory of the Swiss Federal Institute of Technology in Lausanne (EPFL). The aim of the project was the monitoring of civil engineering structures with fiber-optic sensors. This technology was being researched in the aerospace industry for monitoring aircraft, and it seemed interesting to develop the same concepts and techniques for civil structural monitoring. At the beginning, the research team was composed of a physicist (Daniele Inaudi) and a civil engineer (Samuel Vurpillot). This combination allowed them to quickly move from laboratory developments to field tests. The close cooperation with other laboratories within the civil engineering department also proved very beneficial, allowing access to a number of test objects for our technology. Different fiber-optic technologies were investigated, and we finally decided to adopt low-coherence interferometry that seemed to offer important advantages for our application domain, and in particular an excellent long-term stability, a high resolution, and the possibility of creating long-gauge sensors that seemed well suited to the monitoring of large civil structures.

After an initial cooperation with the University of Geneva that provided an existing instrument originally developed for polarization mode dispersion measurements, it became clear that its characteristics were not compatible with the intended use. It was decided to develop a new instrument specifically designed for civil structural monitoring. In parallel, a large effort was put into the development of sensor packaging that would allow reliable installation in concrete or on the surface of existing structures. In 1993, the Surveillance des Ouvrages par Fibres Optiques (SOFO) system, or structural monitoring with optical fibers, was officially born.[1,2] The combination of the new sensors and the new reading unit allowed the first successful deployments in two Swiss bridges and amplified the interest of the industrial partners of the project.

A more ambitious research project was then started in cooperation with the civil engineering company Passera & Pedretti and the Institute of Material Mechanics (IMM). In the framework of this project, Nicoletta Casanova joined the team as the linking person between EPFL and the industrial partners. During that project, we also started to cooperate with DIAMOND, a company specializing in the production of fiber-optic components and, in particular, connectors. Looking for a diversification of their products, DIAMOND was interested in the industrialization and production of the SOFO sensors. This enlarged partnership allowed the realization of large monitoring projects and, in particular, the installation of a large sensor network in the Versoix bridge.

In 1996, the product seemed mature for commercialization, and the young team of engineers met with the industrial partners to see how the SOFO system could be marketed. In the course of this memorable meeting, it appeared that none of the industrial partners had all the competencies necessary to develop and commercialize the system, and it was decided that the creation of a new company would be the best way to proceed. And so, in the course of a few hours, the authors of this article turned from engineers and researchers into start-up entrepreneurs! SMARTEC was founded a few weeks later.

17.2 SMARTEC at Work

In order not to scare away the few readers that survived the historical introduction, it is not our intention to pursue the chronological description of the company development. We would rather concentrate on the strategies that, in our opinion, contributed to the development of SMARTEC.

The strategy of our company could be summarized by the following mottos:

- Money is not enough.
- Be application oriented!
- Put energy into selling your products!
- Offer superior products and support!
- Partner with "complementors."

17.2.1 Money Is Not Enough

Finding investors ready to support a start-up company in the initial stages can be a difficult task, but in some situations having the wrong investor on board can lead to problems worse than having no money at all. A young and relatively inexperienced entrepreneur needs more than money to successfully develop his company. An investor must bring competence to the company in terms of management skills, product development experience, or access to customers and sales channels.

We always thought that our investors must also bring something more than cash into SMARTEC. The initial investors were individuals and companies that could give us management support, guiding us in product development and aiming our energies in the right direction. Being active in a relatively conservative market, it was also important to guarantee that our investors would understand that our business does not explode from one day to the next. In this context, venture investors would not have been a good option. Financing by "business angels" was much more adapted to the company business plan.

We explored cooperation with companies providing complementary technologies. After continuous growth in its first six years, SMARTEC entered a new phase characterized by relative stagnation, with ups and downs from year to year. New business opportunities were identified and explored, but the cash flow from current operations was not sufficient to finance new product- and market-development projects. At that time, we started looking for new investors, and again we focused on companies that could help us further develop our company and products (see "Partnering with Complementors") and on companies providing access to sales channels and customers.

In the end, we found an ideal match in a company that could provide both: the Canadian group Roctest, a public company whose shares are listed in the Toronto stock exchange. Roctest is a world leader in instrumentation and monitoring systems and already owned a division (the FISO Company) providing fiber-optic sensors for industrial applications. In 2006, SMARTEC was acquired by Roctest. This allowed the merger of the sales networks and the combination of the traditional and fiber-optic technologies of Roctest and SMARTEC. The acquisition of SMARTEC positions the Roctest group as the only manufacturer of measuring instruments for the geotechnical and civil engineering markets with a complete set of technologies and solutions to meet the requirements of customers in many different markets. On the other hand, SMARTEC now has access to a larger worldwide distribution network for its current products and has the resources to tackle new markets and developments.

17.2.2 Be Application Oriented!

From its beginning, SMARTEC quickly evolved into a production-engineering company. We realized that many of our customers were looking for a monitoring solution rather than just purchasing components. It also appeared that a lot of support was required during all phases of a project, from system design through implementation to data analysis. Now, our engineering department (called "Solution & Services") can provide complete monitoring proposals that reflect the needs of the customers and include our products as well as components from other manufacturers. From this point, we can offer many options for the implementation of the proposed solution, ranging from turnkey delivery of the system to on-site support from our engineers if the customer decides to install the system independently. These services are offered in cooperation with our local network of distributors and certified "solution providers." SMARTEC has evolved from the "SOFO Company" to a "Structural Health Monitoring Company."

Our application orientation is also reflected in products that are designed for simplicity of use by installers who have no specific background in fiber optics. Besides improving the quality and reliability of the sensors in the demanding environment of a building site, we developed a number of accessories (e.g., connection boxes and installation adaptors) that simplify the use of our components for our customers and for ourselves.

Having installed monitoring systems in more than 200 structures on every continent, we now have strong experience in solving the many practical issues that can make the difference between success and failure. This know-how is transferred to our local solution providers in order to guarantee local support for our customers. On the other hand, we want to obtain as much feedback as possible from any application, so that the experience accumulated locally can be reused in other projects and in other countries. Regular meetings and a central database of applications allows a sharing of experiences among our application partners that strengthen our group after every new project.

17.2.3 Put Energy into Selling Your Products!

SMARTEC has always generated its revenues by selling products. The income from research and development projects represents only a small fraction of our business. This has several consequences:

1. The components that we develop must be highly standardized and reusable in many different applications and situations. Our strength lies in the innovative use of our proven components rather that in the development of custom products that can be used only once. We constantly develop new products (sensors, measurement systems, software packages, complete systems) that complement the existing ones but always aim for a wide use. To date, we have deployed more than 5,000 SOFO sensors and 100 SOFO reading units in more than 200 different applications. Custom developments are carried out only in special cases and when a proven return on investment is demonstrated.
2. The sales team is central to SMARTEC. We rely on professionals with previous experience in international sales of technical products and services. The sales team coordinates the activities of the distributors and

is always looking for partners in new countries. Thanks also to the merger with Roctest, we currently count on a dense sales network covering all relevant countries worldwide. The sales team is also responsible for generating sales in Switzerland and in countries where we are not represented by a distributor. We found that it is very important, even for a small start-up, to count on a strong sales team that is not made from the same people active in product development and production. Otherwise, during busy periods it becomes difficult to insure new orders for the next phase. As we all know, engineers are not always the best salespeople. The sales team can obviously count on strong technical support from the Solution & Services department, and they work hand in hand to develop and sell a proposal suitable for the customer's needs and budget.

3. We strive to develop long-term relationships with our customers and generate repeated sales. This is not always easy, since even a satisfied customer might not have a second application for our systems in the short term. However, giving good support always pays in the long run.
4. In its 10 years of existence, SMARTEC has also diversified in terms of markets, in particular pursuing the oil and gas market in addition to the civil engineering one. The first steps in this market were a reaction to a customer request, but we quickly became more active in developing opportunities in that area. We have found that our core competencies can also be reused in other new fields, provided that we act with the required caution and modesty and that we team up with experts in the field who can guide us in the right direction, both technically and commercially. Examples of other application domains that we are exploring include the automotive and biomedical industries.

17.2.4 Offer Superior Products and Support!

For a company operating from a country with high personnel costs, it is impossible to compete in the global market on the basis of price alone. Therefore, we have positioned our products and services in the higher-end markets, and we look for applications where the superior performance of our sensors offers a real benefit. For example, we can offer unparalleled long-term stability and resolution, and our distributed sensors offer unique performance for long-range measurements. Our customers also expect and get superior technical support during all phases of the project.

Product and service positioning, therefore, play a major role in SMARTEC expansion strategy and in the selection of the technologies to develop and include in our product range.

17.2.5 Partner with "Complementors"

The structural health monitoring market is still in its early stages. At this stage, the main goal for any company should be to develop the market rather than increase its market share. For this reason, we feel that it is in our interest to promote and integrate the products and services of companies that are complementary to ours. A good example is the cooperative agreement between SMARTEC and Oxand in France, a company specializing in risk analysis and lifetime structural management.

On the other hand, we have found that there are cases where our own systems are not suitable for the specific needs of a customer. For that reason, we have decided to distribute and integrate systems from other manufacturers of high-end and innovative monitoring systems. These are usually small and relatively new start-ups that do not affect the efficient, comprehensive distribution network of SMARTEC and Roctest. Also in this case, we can obtain a win-win situation: SMARTEC can increase the efficiency of its sales, being able to respond to more requests with a matching proposal; our partners can concentrate on technology development and technical support and get additional sales from our network. Examples of companies that we cooperate with are Omnisens (a distributed strain and temperature monitoring system based on Brillouin scattering), Sensornet (distributed temperature sensing systems) and Micron Optics (fiber Bragg grating interrogators).

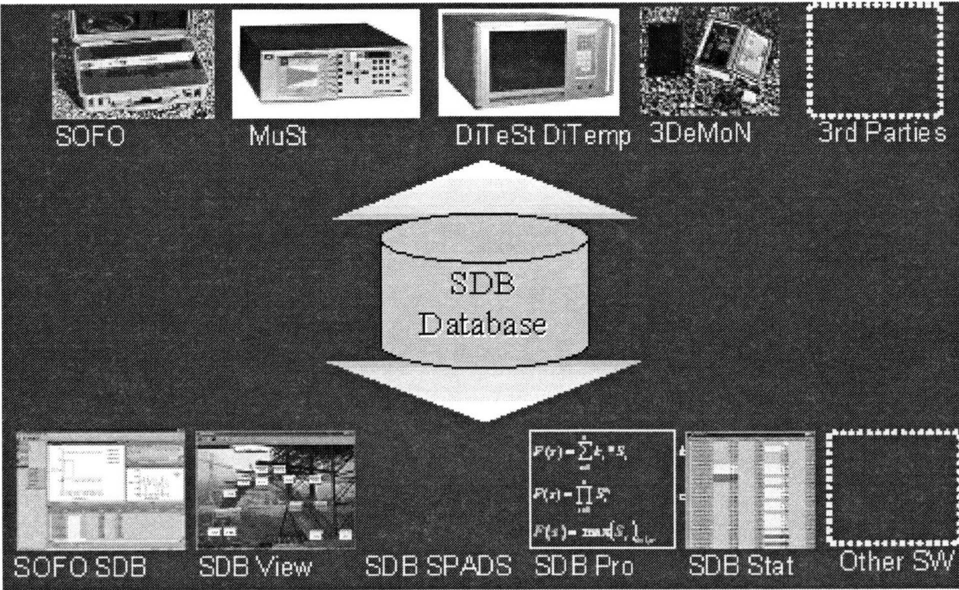

Figure 17.1 The database-centric strategy and monitoring system palette of SMARTEC.

Instead of developing components that are already available on the market, we concentrated on identifying the parts that were missing (e.g., developing special cables and data analysis software that are required for offering a complete solution to our customers). Besides integrating these products at the sales level, we have pursued technical integration, aiming for better interoperability of the systems. There are cases where a number of different systems are required to monitor a given structure. In this case, the client would like to operate all systems from a single software interface and, more importantly, obtain all the measurement results in the same format. For this reason, we have developed a database standard that allows the most disparate monitoring systems to write the measurement results to the same database file.[3] This also allows the partnering companies to develop software packages for data management, visualization, and analysis that are compatible with a wider base of measurement systems, enlarging the potential market. This concept is summarized in the Figure 17.1.

It is expected that these strategic partnerships will further develop and could eventually lead to other mergers and acquisitions. In this regard, a good partnership can constitute a "dating period" for two companies before they decide to get married.

17.3 Conclusions

Fiber-optic sensors have been the springboard for SMARTEC and still generate the largest share of our revenues. Following a strategy based on developing application-oriented systems, putting energy into the sales department, offering high-quality products and services, and partnering with complementary companies, we have evolved from a small start-up to a profitable company employing 15 people. In our expansion, we have aimed for manageable growth that could be sustained by reasonable investments from partners who could also help company development. Despite our small size, we try to apply the principles of larger companies in the development of strategies and partnerships. Bilanz, a Swiss economic journal, has defined SMARTEC as "one of the smallest multinational companies in Switzerland."

The acquisition by the Roctest group opens a new era for SMARTEC and is the natural result of 10 years of pioneering in the application of fiber-optic sensing technology to structural health monitoring.

Acknowledgements

It is obviously impossible to acknowledge individually all the people who have contributed to the development of SMARTEC, so we will do it collectively. We are indebted to all present and past employees of SMARTEC for the passion and efforts that they put into their work, to the shareholders for their continued support, to our partners for fruitful cooperation, and especially to our distributors and customers for believing in the quality and potential of our products and services. We also acknowledge the support of the Swiss confederation through the CTI-Startup initiative and the excellent ongoing cooperation with the Swiss Federal Institute of Technology in Lausanne.

References

1. Elamari, A., Pflug, L., Gisin, N., Breguet, J., Vurpillot, S., "Low-coherence deformation sensors for the monitoring of civil-engineering structures," *Sensor and Actuators A*, Vol. 44, p 125-130 (1994).
2. Inaudi, D., "Field testing and application of fiber-optic displacement sensors in civil structures," *12th International Conference on OFS '97-Optical Fiber Sensors*, *OSA*, Vol. 16, p 596-599 (1997).
3. Vurpillot, S., "Relational database structures for structural monitoring data," *The Present and the Future in Health Monitoring*, Bauhaus-Universität, Weimar, Germany, Edifictio publisher, p 205-214, (2000).

18

"An Earth Odyssey" or Fibersensing

Alberto Maia, Francisco Araújo, José Luís Santos, Luís Ferreira, and Pedro Alves
Cofounders of FiberSensing
Porto, Portugal

Does life sometimes looks like a movie? Or is it the movies that, however fantastic, are never as incredible as real life? Well, the process that led to the implementation in Porto, Portugal, of a technological spin-off on optical fiber sensors with a global vision looks a lot like one of those adventures that we think we can only live in a theater near you.

18.1 "The Meaning of Life" or the Company's Background

At the beginning of the 1980s, the field of optical communications was effervescent, triggering a remarkable set of developments in several science and technological domains such as optical fiber design, modeling and fabrication, compact semiconductor optical sources and detectors, fast optoelectronics, etc.

At the time, the activity in optics at the University of Porto's physics department was essentially focused on imaging and holography. However, the insight of the substantial benefits that would flow from a sustainable R&D effort in optoelectronics and optical communications motivated a young staff to tackle the big challenge of locally developing these areas from an almost zero starting point.

During 1984, contacts were made with the electrical engineering department of the University of Porto to evaluate the feasibility of proposing a large framework project in optical communications. The environment was favorable, considering the political determination of the government to support such initiatives with a strategic decision aimed at technological development in the country, which led to Portugal's integration into the European Union (EU) in January 1986. The project, named SIFO (Optical Fiber Integrated Systems), was approved and launched in May 1985. It provided the logistical and financial framework that allowed the formal concretization of an R&D institute (INESC Porto), mainly owned by the University of Porto, and with a central mission focused on the development of applied research and technology transfer actions.

The SIFO project was very ambitious in its objectives, particularly if compared against the limited knowledge in optical communications mastered by the group at the time. At its peak, around 50 researchers were working on the project, which finished with a successful demonstration in 1990.

As happened in many other places, in Porto the R&D effort in optical communications also stimulated the development of other connected areas such as lasers, optoelectronic instrumentation and integrated optics. In particular, in 1987 António Pereira Leite, from the physics department of the University of Porto, considered that with the technical infrastructure already in place a sustainable R&D program in fiber-optic sensors would be feasible. Abroad, the research activity in this area started in the beginning of the 1980s, triggered by the fast rate of technological developments in optical communications. By that time, as nowadays, the common technological ground between the two subjects was essential to the development of the fiber-optic sensing field.

Local laboratory conditions were at that time already satisfactory to sustain R&D activity in fiber sensing; the need to establish an outside collaboration with an international reference group working in the area was clear. In particular, the possibility of students from Porto pursuing post-graduation programs in such environments was considered critical in order to get a current inside view of the developments in the field. The Applied Optics Group of the University of Kent at Canterbury, UK, headed by David Jackson, was at the top in the fiber-optic sensing area, and António Pereira Leite established contacts with this group aiming for such collaboration. Everything was defined during a visit by David Jackson to Porto in July of 1988, and in September of that year José Luís Santos, at the time a PhD student of the University of Porto's physics department, traveled to Kent, initiating a collaboration that still exists.

In the process of developing the fiber-optic sensing field in Porto, Faramarz Farahi (at the time, senior researcher at the University of Kent and now professor and chair of the Physics and Optical Science Department of the University of North Carolina at Charlotte, USA), was also of vital importance. He has been present for the several phases that framed the local progress of the R&D activity in fiber sensing, helping us with his sharp intuition for novel and important developments and always showing a fascinating sympathetic personality.

In 1992, the Portuguese government launched a large program to carry out a qualitative change of the scientific and technological environment in Portugal. INESC Porto engaged with this initiative and proposed the implementation of a state-of-the-art optoelectronics center. Such a proposal was approved, and during 1992 and 1993 it was possible to see the build-up of an R&D infrastructure with the potential for supporting high-quality work in fiber technology as well as in some areas of integrated optics and thin films. For fiber sensing, this infrastructure had a fundamental role, not only in the expansion of the R&D on multiplexing of intensity and interferometric-based fiber sensors that was already under way, but mostly because it allowed the demand for in-house fiber grating fabrication, a very hot topic by that time. Indeed, in April of 1994, after one year of progress and backward steps, the group succeeded in mastering the FBG fabrication process (see Figure 18.1). With this development, INESC Porto and the University of Porto were in the front line on the R&D involving fiber gratings and their application in sensing.[1] Therefore, the local fiber-optic sensing activity became strongly polarized along this path, a trend that continues now.

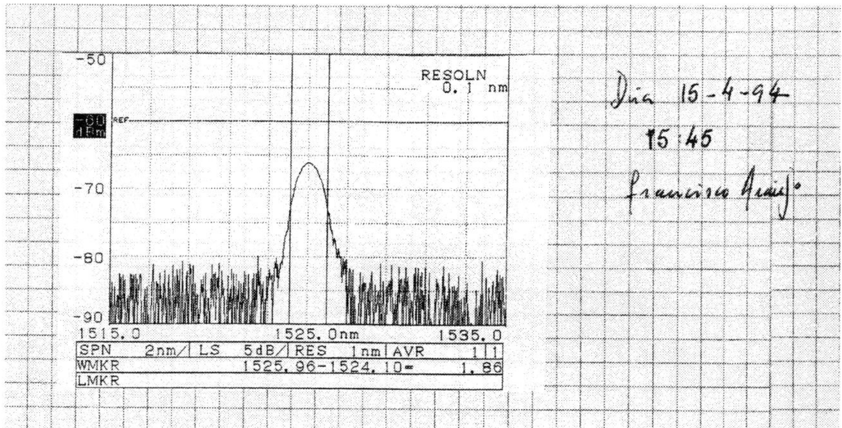

Figure 18.1 Spectrum of the first fiber Bragg grating manufactured at the Optoelectronics Centre of INESC Porto (from Francisco Araújo's logbook).

The second half of the 1990s turned out to be a very fruitful period on the development of the fiber sensing activity in Porto. The implementation of post-graduation programs in collaboration with other universities, the participation in Portuguese- and European-supported R&D projects and networks, the qualitative and quantitative level of the research results obtained—all are indicators of this reality. Also, it was this period in which INESC Porto tried to develop systematic initiatives of technology transfer in fiber sensing through collaborative actions with local companies. However, all of them turned out to be painfully unsuccessful. At the least, they revealed the need for better evaluation of the conditions necessary to efficiently implement this process. This was not a simple job, but one objective feature that helped the analysis of the problem was the recognition of the existence of a large bilateral impedance between the university and industrial environments, despite the reciprocal goodwill. This makes the R&D interaction rather difficult, which is the source of the obstacles.

One possible solution to this situation could be the hiring, by the industrial partners, of people who had undergone post-graduate training, which would help minimize the technology transfer communication impedance. Another possibility, more radical and in line with the Anglo-Saxon approach, could be the concretization of spin-offs from universities and R&D institutes. In this case, there is no impedance matching but, at least at the beginning, it is a more fragile solution since it does not stand in a well-established industrial background.

By 2003 it became clear that, in the Portuguese environment, this spin-off model was the only effective path to transfer the benefits of the R&D in fiber sensing to society. This was indeed the feeling of Luís Alberto and Francisco Araújo, who conceived and set up a plan for an INESC Porto spin-off in this field. In particular, the development of FBG integrated solutions for structural monitoring in civil and geotechnical engineering was central in the technical objectives of the venture. By integrated solutions, it included all of the steps: the conception and fabrication of sensing heads, multiplexing, interrogation units, and software for data processing and display. Other areas of intervention were also considered for a later stage, such as environmental monitoring, aerospace applications, etc. This plan had the necessary institutional support that made it possible to launch the company in April 2004.

18.2 "Goodfellas" or the Founder's Team

FiberSensing's individual founders are José Luís Santos, Luís Ferreira, Francisco Araújo, Alberto Maia and Pedro Alves. The first four came from the INESC Porto environment, while Pedro came from the testing and measurement market.

José Luís Santos did most of his post-graduation studies in fiber sensing in the environment of the Applied Optics Group of the University of Kent, gathering the know-how and attitude that later contributed to consolidating the R&D activity in this area in Porto. Luís Alberto Ferreira and Francisco Araújo did their post-graduate studies in a period where the local infrastructural conditions for fiber technology research improved substantially. Together with their perseverance, this provided a strong background in fiber sensors, particularly in FBG-based sensors, and in measurement techniques, the key point that later enabled the launching of FiberSensing. Alberto Maia brought to the project his expertise in electronics and experience in integrating electronic systems and his team at INESC Porto moved with him to FiberSensing. Pedro Alves was chosen to be the face of FiberSensing. His background in conventional monitoring systems and his management experience, gathered in a few companies in the test and measurement markets, together with his expertise in sales and marketing, made him the perfect piece to complete the puzzle.

José Luís Santos is the only founder who is not connected to the company from a professional point of view. All the others play senior management roles at FiberSensing. This group works very well as a team, complementing each other in an environment of constructive interaction. Each of their professional skills is necessary for the success of the company, and their different personalities ensure that all perspectives of the subject are considered when it is time to make decisions. They're stronger together than they would be individually, and a look at what was achieved since the project started points indeed to the validity of that statement.

18.3 "The Godfather" or Venture-Capital Funding

Poised to be a world-class company in the measurement arena and featuring a very dynamic and talented team, the choice for venture capital as the main financing for the project came very early in the business plan. FiberSensing founders wanted to simultaneously achieve three objectives:

- Set up a highly skilled product development team,
- Develop the first set of industrial prototypes, and
- Establish a worldwide marketing presence.

Being a business with exceptional potential for growth within the $50-billion U.S. target market, with high initial costs involved, few competitors and no historical data, the most logical funding source was an equity exchange.[2] Enter PME Capital, a Portuguese VC, as the business angel who financed the company's first business plan, which was put forward when the company was officially founded in April 2004. INESC Porto secured its share of the company by providing both the team and the necessary technical infrastructure that allowed the beginning of the development phase. This comprised the establishment of well-defined objectives both at the technical level (the development of 10 pre-industrial prototypes and the filing of three patent applications) and at the business level (establishing a presence at international trade shows with potential customer visits and improving the business plan for the industrial phase). The evolution of these two components was continuously analyzed by the company's advisory board, which performed an objective and independent evaluation of the achieved results. FiberSensing's advisory board is composed of seven internationally renowned experts on selected target-market segments and in the fiber-optic technologies and optoelectronics field: David A. Jackson, Professor, Leader of Applied Optics Group, University of Kent, UK; Farhad Ansari, Professor, Department Head, University of Illinois at Chicago, USA; José Miguel Lopez-Higuera, Professor, Head of the Photonics Engineering Group, Universidad de Cantabria, Santander, Spain; Paulo Dias de Carvalho, Engineer, Director EPOS, Teixeira Duarte Group, Lisbon, Portugal; Faramarz Farahi, Professor, Chairperson, University of North Carolina at Charlotte, USA; Francisco Ricardo Nicolas Kaidussis, Engineer, Production Director, Sondagens Rodio, Lisbon, Portugal; and Carlos Dinis da Gama, Professor, President of the Geotechnical Centre, IST-UTL, Lisbon, Portugal.

The founders also contributed to the initial capitalization of the company by acquiring part of the shares. Unfortunately, the Portuguese VC operators do not recognize as an asset the know-how of the individual promoters, the value of an existing team, and having a technologically viable idea that is commercially promising. Therefore, the founders have to put in their own resources to get some participation in the company, which obviously represents a very small share.

The evaluation of the development phase was highly positive and took the company to a new critical moment—its autonomy in human, financial, and infrastructural resources. Autonomy implied a new capitalization stage for FiberSensing, with the corresponding valorisation/dilution process associated with the entrance of a new investor—PME Investimentos, a Portuguese VC firm. After this round, and during the industrialization phase, the company structure was kept as agile as possible and concentrated on activities related to the production and commercialization of its products in order to have enough flexibility to meet different market needs.

In 2004, FiberSensing was granted the Portuguese Innovation Agency (AdI) NEST certification (Novas Empresas de Suporte Tecnológico), which is jointly awarded by the economy, science, and higher education ministries, under the scope of PRIME (the Incentives Program for the Modernization of Economic Activities). This was very important because it provided FiberSensing with access to a special fund, the FSCR PME-IAPMEI (Fundo de Sindicação de Capital de Risco), which is an instrument for risk sharing between operators issued by IAPMEI (Institute for the Support of Small and Medium-sized Enterprises) and currently managed by PME Investimentos. For the founders, this constitutes a potential way of increasing their future participation in the company.

In the same year, the company was also one of the winners of the AdI Ideas Contest (funded by NortInov), whose objective was to sponsor new technological companies based on innovative ideas.

As of November 2006, FiberSensing accounts for €3.3 million of investment, with the major shareholders being the VC firms (PME Capital and PME Investimentos), followed by INESC Porto and the founders. The turnover for 2006 is expected to surpass €0.5 million. For 2007, the main objectives were to increase sales dramatically, add resources to the sales and marketing team, and establish a worldwide network of distributors and certified installers. The sales focus was to be on the applications/industries where FiberSensing's product range has the most competitive position, such as civil and geotechnical engineering and wind generators.

18.4 "The Matrix" or the Product-Development Strategy

From day one, it was a strategic decision to design FiberSensing as a product-development-based company. The master pillar of the company, with its solid R&D background, is the major support of its market positioning: to develop a strong portfolio of off-the-shelf products upon which it is possible to build tailor-made advanced monitoring solutions. These products are sensors, measurement units, and software packages for the implementation of integrated fiber-optic sensing systems.

To prop up the outlined strategy, it was mandatory to establish a multidisciplinary team with vast expertise ranging from fiber-optic technology and optoelectronics to digital electronics, mechanical design, and instrumentation. A major decision was to hire key personnel with diversified industrial backgrounds for each of the critical areas. In a start-up company, it is extremely important to optimize the efficiency of such a team by establishing a set of guidelines that commits each element with the desired end. The paramount goal at FiberSensing is to develop the highest-class FBG-based products in the market, and this underlying rule is ubiquitous to every phase of the product development process—from the concept of the most complex measurement unit to the smallest design detail. FiberSensing's product-development procedure follows the long-established watchmaker's golden principle: for a mechanism to fulfill its function, it is not enough to demonstrate outstanding technical performance, since it is as important to exhibit exceptional aesthetic design.

In order to exploit to the maximum the development effort invested in each product, it was decided to put forward a platform-type architecture to serve as a master for the design of each product. This allows a LEGO-style modular approach for the implementation of highest-order assemblies, with each building block a part of as many products as possible. The development costs are therefore spread over the entire portfolio, and the time to market for new versions or even new products is dramatically reduced. Though the implementation of this procedure may seem straightforward, it is not difficult to imagine that even a limited number of quite simple products are enough to render the complexity of the process unmanageable. In fact, the product-development method implemented at FiberSensing can only be made cost-effective through the adoption of proper product-development support tools, the most important of which is Web-based product lifecycle management software that allows on-demand access to the complete structure of each product. By logging in to the system, it is possible to review detailed technical information on the simplest resistor in an optoelectronic printed circuit board (PCB) or analyze the impact of replacing the thin-film transistor (TFT) screen in the complete series of measurement units.

One rather important edge in the process is to take for granted that although FBG technology is not an obvious solution for all monitoring problems, its inherent advantages enable a straightforward competition with conventional electric instrumentation in many niche applications. This statement, while widening the range of applicability of the engineered solutions, sets intricate constraints on the product-development procedure, the foremost being the market approval of designed products. An important decision was to open the product-development process to reap the benefits of early-stage exposure to market feedback. Letting such constraints rule the process ensures proper leverage for later market competitiveness. The course of action is scheduled in innovation cycles that ensure high efficiency of the process by setting well-established objectives over a restricted period of time, thus keeping the team effort focused.

FiberSensing's innovation cycle is organized in annual periods that are triggered by a meeting of the advisory board to scrutinize novel concepts and new products outlined through interaction with key players in the market. This first trial already provides a high degree of confidence before undertaking a time- and resource-consuming process. The first prototypes are designed by fitting solutions targeted to market guidelines, combining high-quality performance with maximum flexibility towards the implementation of integrated monitoring solutions. Keeping costs under control is also crucial, and it is particularly important to maximize the use of standard telecom optoelectronic components from the earliest stages of product development. Just as important is the implementation of high-precision automation in both the manufacturing and the testing procedures, as well as the continuous training of the manufacturing team (Figure 18.2).

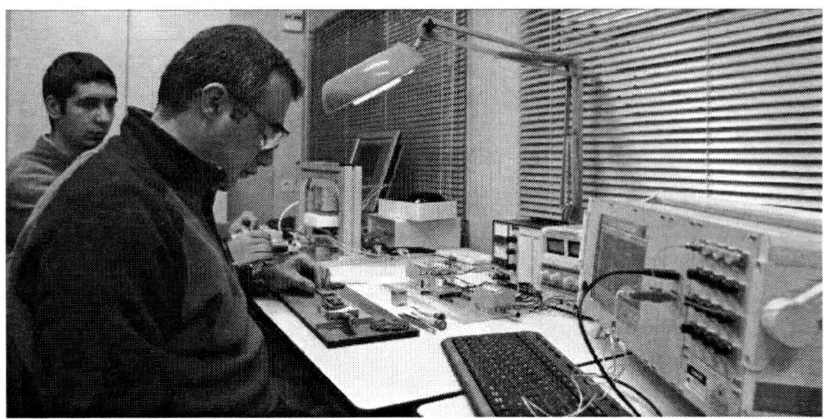

Figure 18.2 Sensor fabrication set-ups at FiberSensing.

Another important aspect is the capital investment necessary to keep the machine running. This revenue is obviously extended over time, and it is mandatory to balance internal resources with complementary financial support. Nowadays, the most important of these sources is subcontracting in the framework of R&D projects, namely the Portuguese Innovation Agency, European Union, and European Space Agency. It is important to emphasize that the extensive competencies of FiberSensing's product-development team are also regularly exploited to perform custom product-development activities through direct contracting by other companies.

It is in this overall context that the intangible asset of the umbilical relationship with INESC Porto gains additional value, since fundamental research assumes a special relevance by closing the innovation cycle related to the generation of new products that ensure long-term market competitiveness.

18.4 "Terminator: Judgment Day" or Market Positioning

One of the most important aspects related to a start-up is its market positioning. Because FiberSensing is a technological spin-off from a research institute, one might think that its market positioning is obvious: highly innovative products that will replace all the conventional options available from traditional competitors. That couldn't be more wrong. In fact, when we talk about fiber-optic sensors for applications such as structural health monitoring, we are talking about a technology that is disruptive for highly conservative markets. Certainly, in areas like civil infrastructures, aeronautics, and oil and gas, it is not enough to present a product and label it as the latest technological gadget. The penetration in these markets implies that the new products are not only advantageous in terms of performance when compared with existing ones, but also, and above all, competitive in price. The typical infrastructure owner will not worry about the technology that will be used to assess the structural integrity; he will simply evaluate the cost/benefit ratio of the proposed solution as a whole. He may even get uncomfortable if he knows that the base technology is new and not one chosen among the ones he has been using for years. Even the engineers that have to effectively use the monitoring solution are more concerned with the application rather than the technology.

These facts were well known by FiberSensing's founders, so a strategy was initially developed to ensure a good, rapid market penetration. Such a strategy was essentially based on the following action vectors: extensive dissemination of the fiber-optic sensing technology, providing the customers with the complete solution, designing the products from a user perspective, competing with both fiber optic and conventional technologies, and establishing key partnerships.

The dissemination of the fiber-optic sensing technology to the potential users of FiberSensing's products was achieved by different means, including

- Direct-marketing near important infrastructure owners and designers, monitoring and analysis laboratories, instrumentation companies, etc.;
- Participation in the most important events around the world related to testing and measurement for different applications, presenting technical papers, and/or having company booths at corresponding technical exhibitions;
- Providing free sensors, equipment, and consulting for demonstrations involving important market players; and
- Implementing R&D projects in concert with potential users of the technology.

This last point is particularly important because it contributes not only to increased knowledge of the advantages of the technology near the application experts but also to product development that is perfectly oriented towards market needs.

One problem with the long time that separated the first demonstrations of the exceptional characteristics of FBGs for sensing and their extensive use in worldwide monitoring applications has been the absence of complete (and comprehensive) solutions in the market. In fact, for a long time companies dedicated to fiber-optic sensing have been providing only scattered products. This means that if an engineer from the application side wanted to use a FBG-based solution, he would have to begin by studying the subject from the optoelectronics side and only then procure providers of FBG sensors (often, bare FBGs), FBG-reading units, and suitable cables. At the end, he may be able to set up something similar to a fiber-optic monitoring system, and probably he would be the only one who could play with it. From this point of view, the idea of providing the customer with a complete solution, from sensors to data-management software, appeared mandatory for the success of FiberSensing. The positioning of the company consisted therefore in delivering a system that could be specifically tailored to fulfill all the requirements of the customer's application in a comprehensible way, so that the technicians who have to deal with it daily, and who usually are only familiar with conventional electric-based instrumentation, can take real advantage of the technology.

This relates to another point: competition. It is well known that the sensor market is huge.[3] So theoretically, there is space for a large number of fiber-optic sensing companies.[4,5] The problem is that fiber-optic sensing companies have to compete with well-established companies operating with traditional technologies, which do the job at acceptable prices in a large range of applications. Of course, there are a few applications where only fiber-optic sensors can be utilized. But if one is limited to those applications, the size of the market is drastically reduced. Consequently, product designs must take into account three essential aspects:

- The cost of the complete solution. The price per sensor must be comparable with the ones from competing technologies.
- Simplicity. Sensors must be provided with connectors, eliminating the need for splicing; measurement units must allow fast, simple sensor configuration and must have an intuitive, user-friendly interface.
- Comparison with similar existing products. For example, strain sensors embedded in concrete must be similar to vibrating-wire sensors currently available for the same application so that they become recognizable by typical users.

Of course, other essential technological improvements were pursued to ensure a successful penetration of the FBG-based sensing solutions on the monitoring market. From the point of view of the sensors, these included increasing the number of parameters that can be measured, providing packaging oriented to a variety of applications, and guaranteeing long-term reliability. With regard to the measurement units, important drivers were increasing the number of interrogated sensors, increasing dynamic ranges and sampling rates, adding more compactness and portability, and including data-logging and data-transmission capabilities as well as hybrid capacity (mentioned below).

In the future, standards definition in FBG-based sensing systems will also be important. Up to now, only a few isolated efforts in this direction are known, but certainly this will be an important movement for this technology. [5,6,7]

The disruptive nature of the fiber-optic sensing technology was clearly identified by the FiberSensing executive team as one of its most important drawbacks. The idea that users will employ FBG-based sensors in all their applications is far too optimistic. To overcome this, FiberSensing introduced the hybrid platform, which allows the simultaneous assessment of signals generated by a sensing network composed by both electrical and FBG sensors. Based on this platform, different measurement units have been implemented, which are now commercially available. These units have a graphical interface that allows the user to configure optical and electrical sensors in a similar way, as well as acquire, operate, and store both types of signals. The units are expansible (i.e., it is possible to increase the number of electrical and optical channels by inserting additional optical-switching modules or data-acquisition cards). These characteristics can increase the fiber-optic sensors' opportunities on today's most demanding applications, where they still need to be complemented with conventional electric sensors that are either cheaper or already installed on site. To further explore this idea, FBG interrogation modules were also developed to be used in standard data-acquisition systems, already present in most test and measurement laboratories. This is the case of the PXI BraggScope, which allows the owner of any PXI controller to use FBG technology by simply installing a two-slot PXI card module and a device driver.

Finally, the partnership issue. INESC Porto is an obvious partner for FiberSensing. By promoting fundamental research activities on multiple fiber optic sensing areas, it is a constant source of new ideas with potential commercial interest. Once again, this may ensure innovation, but market penetration is a different matter. From this viewpoint, it is particularly important to establish partnerships with key companies or entities that position themselves near the market (i.e., near the applications), and that can benefit from using FiberSensing technology. By developing complete solutions driven by specific applications, it is possible to obtain competitive monitoring systems both in terms of performance and cost, allowing those partners to win new customers and FiberSensing to assess markets that otherwise would be very difficult to reach.

To be continued... ("We'll be back!").

References

1. López-Higuera, J.M., Ed., *Handbook of Optical Fibre Sensing Technology*, John Wiley & Sons, Chichester, UK (2002).
2. Intechno Consulting, *Sensor Markets 2008: Worldwide Analyses and Forecasts for the Sensor Markets until 2008*, Basle, Switzerland (May 1999).
3. Rzhavin, Yu. I., "Fibre-optic sensors: technical and market trends," *Measurement Techniques*, Vol. 46, Pg. 949 (October 2003).
4. Krohn, D., "Market opportunities and standards activities for optical fiber sensors," *18th International Conference on Optical Fiber Sensors*, Cancun, Mexico (23-27 October 2006).
5. Riviera, E., Mufti, A. A., Thomson, D. J., "Civionics specifications for fibre optic sensors for structural health monitoring," *Proceedings of the Second International Workshop on Structural Health Monitoring of Innovative Civil Engineering Structures*, Mufti, A., Ansari, F., eds., ISIS Canada Corporation, Winnipeg, Manitoba, Canada (2004).
6. Udd, E., Inaudi, D., Culshaw, B., Ecke, W., "Fibre optic sensor opportunities and obstacles for aerospace and civil structure applications," *Panel Discussion in Smart Sensor Technology and*

Measurement Systems–SPIE Conference 5384, Smart Structures/NDE 2004, San Diego, California (2004).

7. Brönnimann, R., Held, M., Nellan, P. M., "Reliability, standardization and validation of optical fibre sensors," *18th International Conference on Optical Fibre Sensors*, Cancun, Mexico (23-27 October 2006).

19

The First Years of Crystal Fibre A/S from a University Perspective

Anders Bjarklev and Jes Broeng
Cofounders of Crystal Fibre A/S
Denmark

This chapter describes some of the essential steps towards the formation of Crystal Fibre A/S as seen from the perspective of the employees of the co-founding university, the Technical University of Denmark (DTU). The chapter addresses issues concerning barriers for the development and important decisions leading to the founding and first years of the company history.

19.1 Introduction

More than a decade has passed since we took the first steps towards the formation of Crystal Fibre A/S, which develops and manufactures photonic crystal fibers. These optical waveguides, which also are known as microstructured optical fibers, in most cases confine light to their central part by a highly organized structure of microscopic air holes running along the full length of the fibers. A lot of knowledge has, of course, been gained since the beginning of this technological voyage from a scientific perspective, but also from the perspective of bringing technology beyond the laboratories of the university. Surely, this process will also look different depending on the role that one plays in it, and it is, therefore, very important to have in mind that the experience to be gained from other people will be strongly colored by their positions.

A very important subject in the public debate concerning the innovation process is the role of different institutions within societies, and a strong interest from a political point of view is the question of the mutual roles that should be played by private enterprises and publicly funded institutions. What each person believes to be right and wrong in this discussion is basically a question of personal and political standpoints, but independent of this, it is relevant to learn from experience under all circumstances. The present contribution should be seen in this light, taking as a point of reference the role played by a university (and its employees) in the process of forming a novel, independent company in collaboration with private industry.

19.2 The First Years

When we look at the process of innovation, the first ideas are, of course, very important. It is interesting that when you try to recall the first steps in this process, you will often be strongly biased by the knowledge that you gained through the process. Furthermore, I believe that the search for new ideas has some of the same elements as the search for other and more specific things—for example, when you are looking for a specific tool in your workshop or your favorite knife in the kitchen—namely that we are looking in circles.

With the knife, for instance, assume that you haven't used it for several weeks. You may have an idea where it was placed, and that is probably where you start to look, but if unsuccessful you will extend the search until you finally find the knife (or choose another one). When you are asked where you found the knife, you will only point to a certain drawer and not talk about the fact that you had to go through the whole kitchen to find it. I believe it is the same with research subjects. In both cases, we search in a more or less organized manner, and with more or less dedication, but whether you find what you are looking for strongly depends on the approach. I believe it is essential that the overall goal is clear (otherwise, we get distracted) and that the working method is guided by a systematic and careful search.

In the case of photonic crystal waveguide research, we were strongly inspired by the early work (1987) of Yablonovitch and John explaining the ideas of controlling light through the use of periodic media, and this inspiration was further fuelled by a lecture by Yablonovitch in Copenhagen in 1993.[1,2] From our point of view, the driving force at this time was that we had seen a novel way of controlling wave propagation (in the microwave region), and we felt completely sure that this fabulous tool would provide advantages for optical wave propagation and control. To continue the knife-in-the-kitchen analogy, we did not exactly know whether we would find a knife, fork, or spoon, but we felt certain that it would be extremely useful. As it turned out, we found a multipurpose tool that was hidden away.

Our approach was to define an initial project on multidimensional periodic structures, and with strong support for this from our university management, the first PhD project was defined in the early part of 1996.

The subject of the initial project was quite broad, and our first actions were, therefore, to focus the efforts. From multidimensional optical structures, the goal was narrowed to photonic crystal fiber structures. We considered the manufacturing of a two-dimensional optical structure more accessible than other suggestions, and we had strong experience in optical fibers at the Technical University of Denmark. We were, however, aware that a practical demonstration of the photonic bandgap effect, which was considered the holy grail of the optical community that formed around Yablonovitch's and John's ideas and was propelled by the research group of Russell at the University of Bath, UK, would lead to more general acceptance of the new ideas.[3] This issue must be seen in light of the general scientific interests of the time, where the subject of photonic crystals certainly was not mainstream.

Figure 19.1 Micrograph of the cross-section of a typical photonic crystal fiber from the early years of Crystal Fibre A/S. The black areas are holes.

19.3 Scientific Ideas Start to Take Shape

In the initial phase of the photonic crystal fiber project, the basic understanding had to be formed, and we needed to establish reliable ways of testing new ideas. We took our point of reference from a well-established standpoint, namely numerical modeling of optical waveguides, which in previous years had been an essential tool in the development of high-performance, erbium-doped fiber amplifiers (see, e.g., Bjarklev).[4] We therefore set out to establish a full-vectorial electromagnetic model for analyzing and (together with proper optimization tools) synthesizing optical fibers with periodic air holes placed along their length. In mid-1997, this was a valuable step with respect to understanding the then-surprising properties of this new class of waveguides, and the improved understanding was, of course, a key element in the growing ideas for patenting our findings.

An important element in understanding our actions at this point is that we (like most researchers) had to publish our results to qualify for funding for future projects, and the new PhD project also had to show progress with respect to publication. At the same time, we started to discuss the idea of patenting, which due to the law at the time was left to the researchers' own initiative (this was later changed by a 2000 Danish law formalizing the more active role of the universities). Some national programs supported patenting by providing advice and courses to interested researchers, but economic support was highly limited at the time, and experience and recognition within the scientific society for these activities was limited.

On the purely scientific side, we were in a good position at the end of 1997, forming an early collaboration with the University of Bath, UK, which had fabricated the first microstructured optical fibers during 1996 and now were on the search for the experimental demonstration of the photonic bandgap effect. The collaboration resulted in an agreement, where Jes Broeng, with the latest numerical predictions from DTU, traveled to Bath to join Jonathan Knight and his coworkers in fabricating a photonic bandgap fiber.

19.4 External Funding Is Obtained and Breakthroughs Finally Happen

During the early days of 1998, our expectations of the planned collaboration with the University of Bath were quite high because we had seen the first indications of strongly located optical fields in honeycomb-structured fibers having holes of the rather limited size that was possible at the time—and we knew that the Bath group was highly skilled with respect to fabrication. This point is also very important because it held the central, driving force for us: we believed we had seen the first (at this point theoretical) evidence of the photonic bandgap effect.

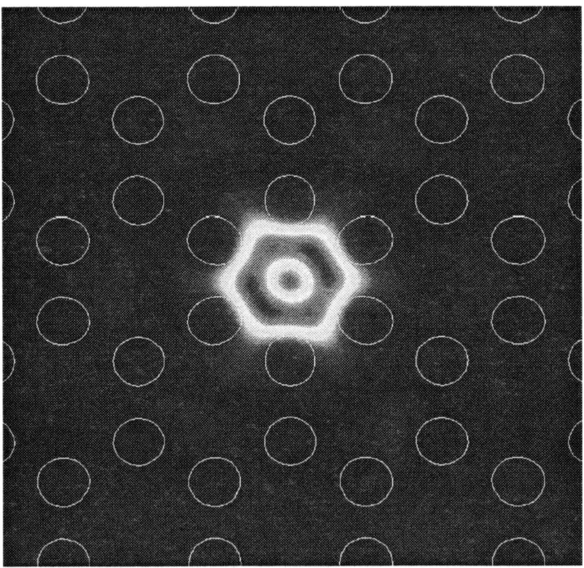

Figure 19.2 The first numerical predictions of located optical fields in honeycomb structured fibers. The calculations were made at the Technical University of Denmark during 1998.

Another important step was that our National Research Council decided to grant support for our project, and our local management was fully behind the project even though we had been unable to publish the results of our efforts. This was probably also related to the fact that the optical-fiber community was working very hard to meet the demand from the booming telecom industry, and that the standard fiber technology was doing fine, so the need for a completely different approach was not in the minds of many reviewers. The funding for patent work was still hard to find, so we decided to draw on our own and internal experiences within the university to establish the first protection of our intellectual property rights (IPR) before the collaboration with external partners could really gain speed.

I would like to stress that I find it extremely important that the national and international research-funding bodies take the necessary risk in the early phases of new research areas—at least when it comes to research that has its origin in the university environment. Also, in this phase it is of the highest value that the local management is behind the researchers and that they create the necessary research space. These elements were essential for our case as well for other colleagues.

In the final part of 1998, Jes Broeng went to the University of Bath, and only three weeks after he left with the fiber design in his briefcase, we received a phone call from the laboratories in Bath. Everyone was extremely happy because they had seen the first actual evidence of the photonic bandgap effect with their own eyes—a very bright, colored mode field, which could be clearly seen in the microscope. This result, which was published in *Science* shortly after, formed a real breakthrough from a scientific point of view because all the resistance from reviewers and journal editors seemed to be replaced by a much more positive attitude towards the idea of the photonic bandgap effect.[5]

19.5 The Formation of Crystal Fibre A/S before the New Millennium

With Jes Broeng coming back from the United Kingdom with fresh results around Christmas time 1998, we were in a completely different situation, and we were highly focused on taking the scientific success even further and aiming for commercial activity.

In this process, we found some help from a new side. The Danish weekly journal *Ingeniøren*, which is read by most Danish engineers, took up the story, and an article printed in the early days of 1999 became helpful in spreading the news beyond the readers of *Science*. Also, John Heebøl, the director of a publicly funded organization called University Innovation, saw the perspective in the new idea and chose to support the patenting process, which allowed the involvement of professional patent attorneys.

The aim was now to seek the necessary funding for a start-up, and we decided to join forces with the well-established Danish industry group NKT, which had shown a remarkable track record concerning high-tech start-ups and had collaborated with university partners for years. This step involved some new elements compared to the normal life of university researchers because the whole line of questions concerning legal matters became much more important. From the researcher's side, we had a clear focus from the beginning to keep the university involved in the process, although the law at that time did not require this. We were, in other words, free to act on our own initiative, but we also saw some of the drawbacks of this in previous cases. We really wanted to avoid a situation in which we would have to take sides between the new company and the university. Our viewpoint was the following: the photonic crystal fiber technology still needed to be strongly coupled with the know-how of the university, and the only manner in which the interests of a private company and the university could be coordinated was through a mutual agreement.

19.6 The Company Grows with a Clear Focus on the Customers

The negotiations between NKT and the Technical University of Denmark had been completed with a so-called "field of collaboration," which stated that research done within the area of microstructured fibers at the university would be offered to the newly formed company, and in return the university received warrants (which later were called and the shares received for the delivered research results in the form of intellectual property rights). It was very important to make this specification up front because it allowed the involved researchers to see what elements of their work had/had not been included in the collaboration, and it clarified that activities within these areas could be done in the interest of the university as well as the company.

The first year of the company's life was naturally a build-up phase, where the company rented offices, laboratories, and (maybe most important) know-how at the university. During this process, the industrial partner brought in essential market knowledge and especially a clear focus on customers, who initially were research laboratories interested in access to the new kind of optical fiber. During this period, there naturally was daily collaboration between the employees of the company and those of the university, with a strong feeling of pulling in the same direction concerning scientific results. However, at the same time we were all very clear on the different roles of the partners, and it was decided from the beginning that the commercial aspects (such as customer contact, prices, etc.) should be handled solely by the company. The role of the university employees was to create new scientific results and to formulate these in papers and IPR.

One of the most important issues at this stage was the choice of a personal role in the new company versus the university. Personally, I find that motivation remains at a high level only if the involved people have clarified for themselves and their families what they should aim at. Not doing so may cause serious confusion if the individuals one day act as a company person and the next as a university employee.

It is very important to note that although this period required the closest possible collaboration between the two groups, the clear definition and mutual understanding of roles allowed us to achieve numerous results for the company as well as the university. Crystal Fibre A/S developed new fiber products, and the university was able to write many new scientific papers and educate new students within the emerging field of photonic crystal fibers.

19.7 A More Mature Company Emerges and the University's Role Changes

After less than two years in rented facilities within the Technical University of Denmark, Crystal Fibre A/S moved to its own premises in Birkerød, about 15 kilometers north of the university. This allowed the company to establish full-scale production facilities but, at the same time, keep close contact and collaboration with researchers employed at the university.

The collaboration between the young company and the university was strengthened by carrying out a number of industrial PhD projects, but it was also an important time with respect to the further definition of the partners' new roles. The collaboration became more and more similar to those of any other interaction between the university and a private company. As Crystal Fibre A/S developed, NKT decided to buy the UK-based company Blaze Photonics, which since its foundation by the University of Bath group had been a competitor of Crystal Fibre A/S. With this fusion, the know-how and IPR bases of the two companies were joined, forming what today is the world's leading manufacturer of photonic crystal fibers.

19.8 Conclusions

From the perspective of a university professor, I would like to say that the first years of Crystal Fibre A/S have shown the importance of having university management with a clear focus on innovation and a clearly supportive attitude towards the initiatives of its employees. It is essential that projects can be discussed openly with the university administration and that the common knowledge of development may be improved from case to case.

I would also like to say that a lot of persistence and hard work has been necessary all along the way. There have been many situations where things looked difficult, but we continued simply because we believed that the ideas were good enough, and because the work was fully supported by our organization.

It also has been very important to seek advice from colleagues and from experts in other fields. However, this is probably one of the most rewarding elements of the process because this allows you to interact with many new and exciting people.

Finally, I think that it is absolutely essential to be clear on your personal ambitions and to find a good collaborating partner with capacities and visions that complement your own. This is a very difficult task because the specific case depends on a lengthy development process in a market that very few people can predict.

Under all circumstances, it is great to learn new things every day of the process!

References

1. Yablonovitch, E., "Inhibited spontaneous emission in solid-state physics and electronics," *Physical Review Letters*, Vol. 58, pp. 2059-2062 (May 1987).
2. John, S., "Strong localization of photons in certain disordered dielectric superlattices," *Physical Review Letters*, Vol. 58, No.23, pp. 2486-2489 (1987).
3. Knight, J., Birks, T., Atkin, D., and Russell, P., "Pure silica single-mode fibre with hexagonal photonic crystal cladding," *Optical Fiber Communication Conference*, Vol. 2, p. CH35901 (1996).
4. Bjarklev, A., *Optical Fiber Amplifiers: Design and System Applications*, Artech House, Boston-London (August 1993).
5. Knight, J., Broeng, J., Birks, T., and Russell, P., "Photonic band gap guidance in optical fibers," *Science*, Vol. 282, pp. 1476-1478 (Nov. 20, 1998).

20

Multiwave Photonics: Building a Fiber Optics Company in Portugal

Jose R. Salcedo
Professor, Founder and CEO of Multiwave Photonics
Porto, Portugal

The author has an impressive track record in Portugal as an academic, entrepreneur, influential civil servant, and government advisor. Here he brings all this together in the short but highly telling story of Multiwave Photonics, highlighting not only the need for business and markets but also the variation among countries and cultures.

20.1 Introduction

Multiwave Photonics is a high-tech start-up company based in the historic city of Porto, located in northern Portugal and bordering the Atlantic Ocean in the southwest of Europe. The company is focused on optical sources based on fiber-optic technologies and predominantly targets OEM business opportunities. The key target markets are international in nature and are associated with selected industrial applications and medical imaging applications.

Multiwave Photonics' products are normally provided as compact, cost-effective OEM modules but can also be provided as small benchtop instruments. They include pulsed seed lasers and master oscillator power amplifier (MOPA) fiber lasers for selected industrial applications, incoherent optical sources (ranging from narrowband sources that can replace lasers in selected sensing and telecom applications to broadband sources for medical imaging [optical coherent tomography] and instrumentation applications), specialty optical amplifiers such as single polarization and polarization-maintaining ytterbium- and erbium-doped fiber amplifiers (YDFAs and EFDAs), and subsystems that incorporate some of the previous products and core technologies, such as fiber sensing temperature and vibration monitoring systems for electrical high-power transformers and mechanical structures that include optimized multichannel optical sensor interrogation units.

20.2 Motivation

The motivation to create a company related to lasers and fiber optics was developed during my graduate-student years at Stanford University in the late 1970s, where I was exposed to the entrepreneurial spirit of Silicon Valley through the outstanding scientific and entrepreneurial environment at the Ginzton Laboratory in particular and Stanford University in general. However, creating a Silicon Valley-like company in a more traditional society poses particular challenges because the basic conditions easily found in Silicon Valley are only now becoming available to some extent in Portugal. In fact, Silicon Valley is a unique melting pot of competencies, venture capital, entrepreneurship, risk-taking culture, and international networking that is difficult to reproduce elsewhere. And, to reproduce it elsewhere, key cultural factors must be in place.

Looking at these conditions in turn, competencies are now much more widely available in many scientific areas in Portugal, after years of continuing public investment in universities and R&D institutions with particular emphasis on sponsoring PhD fellowships and R&D programs in the country and abroad. For example, fiber optics developed locally over the past 25 years, and this was a process where we played an important role and that later helped to anchor Multiwave. Similar positive evolutions occurred in other science and technology (S&T) areas, in particular in the biomedical sector, where the country now has excellent S&T institutions and world-class competencies that are also helping anchor additional start-up companies.

20.3 Area Social Factors and Financial Support

Capital to support start-ups or high-tech ventures, however, is almost not available in Portugal, at least not in the usual sense of venture capital. Local so-called risk-capital firms often manage public funds and have little or no experience in the private sector. When they have, though, such experience tends to be focused on medium-sized companies where government and public-minded institutions still carry a significant influence. Private funds do exist, but they tend to avoid earlier-stage and higher-tech operations due to the perceived higher risk, preferring to focus on development capital, later stage financing, and mergers and acquisitions (M&A).

In very recent years, entrepreneurship has entered the political vocabulary, and significant public funds started being allocated to promote new enterprises. However, prevailing attitudes did not change significantly, as those funds are often managed in the older ways: political sensitivities play an important role, investment decisions are often taken without any significant due diligence, intellectual property rights are not clarified at the onset, and exclusive commitment of the founding team is considered too risky and even alien to become a basic requirement. Even with these limitations, the situation has vastly improved in recent years.

From our experience, the greater difficulties in a more traditional society lie with prevailing attitudes, and these are strongly determined by social factors accumulated over an extended period of time. In fact, people tend to avoid risks, as failure usually carries a negative social stigma and heavy shame-related penalties. In addition, people tend to isolate themselves in noncommunicating subsystems: students barely interact with teachers, scientists do not normally interact with investors, and universities barely interact with companies. Unfortunately, such cultural and social behaviors introduce additional difficulties for entrepreneurs, especially internationally minded entrepreneurs who may need more capital or face more complex risks to compete in international markets.

Our own experience substantiates these perceptions. We initially funded our original Multiwave Networks company in the Northern California VC market, raising $22 million in May 2002. Just one year later, and after having invested about $10 million in the company, the telecom market collapse provided us with an opportunity (and corresponding challenge) to purchase the company from the original VC firms in a management buyout operation. We spent the following year restructuring the company and refocusing its business plan, and thus Multiwave Photonics was born. We then looked for financing above and beyond the financing we were already providing ourselves.

For reasons that we perceive as intrinsically cultural and social in nature, financing Multiwave Photonics in Portugal proved too difficult. Fortunately, in nearby Spain we found Bullnet Capital, an excellent institutional investor who understood the company and became our partner, closing a funding round in April 2005.

Creating, financing, and developing an internationally minded high-tech start-up company in a traditional society usually requires dealing with and navigating through complex cultural and social issues. In our case, we decided to bypass them as much as possible and follow typical Silicon Valley-type procedures, including building an international team, filing U.S. and international patents, developing state-of-the-art and internationally competitive products based in fiber-optic technologies, following well-established and demanding international due diligence and advisory processes, and making the company a normal participant of international networks related to lasers and fiber optics. In this context, we sought and are fortunate to have earned the support of a number of distinguished advisors who make up our technical and business advisory boards. Our advisors are exceptional individuals, and many became early angel investors, having supported the team and company since day one. They provide us sharp advice and criticism as well as helping open new opportunities, including the identification and building of strategic relationships in areas where we find appropriate synergies.

Multiwave Photonics is focused on a particular business (optical sources), but the challenges the company has faced seem fairly general in nature and could well apply to any high-tech start-up company based anywhere. In fact, you need to combine a wide spectrum of ingredients, including capital, hard work, razor-sharp focus on specific market opportunities with products that are good enough, networking capacity, the ability to learn and adapt, the ability to say no, constant attention to people, and a brutally honest approach to reality. In addition to these, you may also need a few other ingredients to survive in and adapt to local circumstances; these other ingredients are mostly cultural and social in nature, and social skills come in handy. They do not resolve the issues, but they may help you understand them, a not-so-easy task anyway.

20.4 Conclusion

Europe (not just Portugal) will benefit greatly by adopting better-informed, professionally managed, risk-taking programs, criteria, and attitudes to sponsor innovation and entrepreneurship. It will take a few years to clean the system of negative bureaucratic influences, but it must and will happen. The prevailing negative feedback mechanisms in society must gradually be replaced by positive feedback with some control. In the meanwhile, many younger and not-so-young people will be busy creating and developing innovative companies, building the future in which they believe and showing that, yes, it is possible, worthwhile, and fun.

Part III

Supporting the Entrepreneur

The main aim of this section is to present some key contributions to support the companies: the intellectual property rights, the venture capital, and trade organizations.

Every embryonic small company and most of the big ones need partners and support. A quite amazing plethora of organizations have sprung up to provide this support.

Some are in themselves independent commercial entities—the lawyers, bankers, and accountants. Sometimes our entrepreneur seeks input from the management consultants, market researchers, and business-development planners, all of whom in one form or another work for a fee to provide necessary input. Many of these are generic and can equally apply their wares to any part of the small-business sector. Some, though, are very pertinent to the technology-oriented entrepreneur. In particular, we have included venture capital and intellectual property law as two of our three chapters in this section.

It is in the nature of a product-oriented technology venture that extended product development is almost inevitable. This includes not only the realization of the product itself but also the initial marketing infrastructure, covering the often-required legal and regulatory framework demanded by the potential customers. These processes are invariably demanding in both time and manpower and, of course, earn nothing in the early phases. For the small company, significant external capital is frequently the only route. Indeed, evolving beyond the tiny—perhaps 10 or a dozen employees—almost inevitably leads to the offices of venture capital.

Our example here, Bullnet Capital in Madrid, portrays the activities of a small, specialized funding group targeting high-technology ventures within the cultural context of Southern Europe. They specialize in the relatively modest sums that are often difficult to obtain from the large houses and advance as appropriate into further higher-level funding partnerships as the project develops.

There are countless other venture-capital funds, often administered by a small group of individuals and often specializing in specific areas. Many successful entrepreneurs have also spent time on this side of the fence—and Stuart Barnes, whose contribution appears in the first section, is one such example. Venture capital is loved and loathed in equal measure within the entrepreneurial community, but it is undoubtedly a necessity.

Intellectual property and its protection are invariably portrayed as another "desirable," and our guide to IPR law, based in the United States, conveys this general theme effectively. There are, though, hints of realism to be applied to IPR and all it entails. Perhaps the most important is that most small companies cannot even contemplate patent litigation. The function of patents is then at best to provide measurable (though sometimes of questionable meaning) comfort to investors. Furthermore, most technological patents are highly specialized and, perhaps paradoxically, easily circumvented. If the motivation is to prevent others doing the same deed, then publish in a journal with a sufficiently low readership. In other words, patenting has a role, but it in our local experience has been something to avoid (or at best to invest in with great caution). Recently, we have also encountered several members of the venture-capital community who are of similar disposition. Agility, corporate loyalty, and ingenuity, while less easy to measure, are significantly more useful assets.

The third of our supporting organizations represents the trade associations that provide a meeting ground for like-minded individuals with similar issues and also stimulate exchanges into other communities. The importance of these organizations is both hard to quantify and difficult to overstate. Trade associations build an entrepreneurial community. They provide critical mass for lobbying government and major organizations such as utilities and large corporate entities. They will negotiate exhibit space and attract professional society involvement and, therefore, conferences and communities from an international dimension into a local context. They are the all-essential catalyst for community advancement.

21

Developing High-Tech Companies in Spain and Portugal

Javier Ulecia
Founding Partner and CEO
Madrid, Spain

Bullnet Capital portrays the activities of a small specialized funding group targeting high-technology ventures within the cultural context of Southern Europe. We specialize in the relatively modest sums that are often difficult to obtain from the large houses and advance as appropriate into further, higher-level funding partnerships as the project develops.

21.1 Background

Spain at the end of the 1990s, like all other developed countries, was frantically living the Internet bubble. All business projects, qualifying as "technological" merely because they employed the opportunities that, at this moment in time, the Web had to offer as a new communication and distribution channel, encountered without much difficulty the financial and human resources required to start up. Professionals with diverse backgrounds and experience abandoned their jobs, many well-paid and with a promising career ahead of them, to become entrepreneurs seeking quick and easy success. On the other side of the equation, the fund providers placed at the disposal of these new entrepreneurs immense quantities of capital in the hope of benefiting from returns never before achieved in such a short period of time. To this end, innumerable new investment vehicles were created under the name of incubators, catalysts, or accelerators, converting themselves into icons of the New Technological Era.

However, before the arrival of Internet Fever, and unlike countries with a greater entrepreneurial culture such as the USA or the UK, in Spain and Portugal it was practically a miracle that innovative business projects emerged with a leading technological component, since neither the culture nor the infrastructure existed: the "technological" community was averse to taking any risks; incentives from sources (such as investigation centers, universities, etc.) were lacking; very few private companies were able to sustain the development of leading-edge technology; it was very difficult for new projects to access capital; there was very little support from public authorities; and, in short, there were very few success stories.

Did the bubble help the development of said culture and infrastructure? Not really. The projects emerging during the heat of the Internet were, in the vast majority, were business models that used existing technology but did not develop innovative technology. Also, the investors supporting said projects were, in general, financial and administrative in nature and had no experience in the world of technology, the opportunities available, nor its limitations. More important, during the year in which the bubble burst, the majority of the companies, investment vehicles, and other protagonists in this sector disappeared as quickly as they had appeared, with the subsequent negative impact on the already damaged reputation of technological innovation. Not only were there very few success stories but many much-talked-of failures.

In short, the main causes behind such collective failure can be summarized as follows:

- Investments made purely to take advantage of the momentum and not founded on any basic (or, in many cases, logical) criteria
- Completely outrageous valuations
- Saturation of emerging sectors: too many start-ups in markets whose existence was yet to be proven
- Confusion between investment in differential technology and "Web boutiques"
- Lack of focus, with many investors abandoning their natural field of expertise
- Very short-term project vision

21.2 Genesis of the Company

Around the middle of 2001, at the height of the aftermath of the bursting bubble, Bullnet Capital began to take shape. It would be logical to think that the market situation was not at its most favorable for launching a new capital venture fund specializing in technology; but our hunch was that, despite the bleak panorama, the sector could generate attractive investment opportunities, and it was the ideal moment to sow seeds in solid projects. Moreover, and given the very little interest by the investment community in technological projects, we thought we could dedicate sufficient time to analyzing investment opportunities and, at the same time, increase our negotiation power and by default improve our entry conditions.

It must be said that the three founders of Bullnet Capital (Bruno Entrecanales, Miguel del Cañizo, and He-Who-Writes-These-Lines), were not only witnesses but also protagonists in what happened during those years. In our role as assessors, consultants, or investors, we were partly responsible for what occurred. Despite this, we were (and still are) completely convinced that Spain (and

Portugal) were still ripe for the picking, and that sooner or later the conditions would arise for truly innovative technological projects to emerge. For this reason, while trying to learn not only from our own errors but also those of others, we decided to swim against the current and establish our fund based on several basic arguments:

- Focus on projects related to information technology
- Product positioning (vs. services)
- Project selection with a technological differential value sustainable over time
- Instinctive drive to exit even before investing
- Take every caution taken during the due-diligence process, knowing how to combine technological and strategic analysis
- Add value and contribute to the day-to-day running of participating companies and be informed of the key business issues

Helped to a great extent by the initial support of Bruno Entrecanales, financial sponsor of the project, we managed to convince a series of private investors as well as an institutional investor, the European Investment Fund, to put together an initial fund of €18 million in March 2003 to support the launch of technological start-ups in Spain and Portugal. Our initial objective was to invest in five or six companies about €2.5 million each over the following five years.

Investment criteria were (and still are) as follows:

- Companies with business models differentiated by
    - Long-term sustainable opportunities with high potential growth rate
    - Led by highly motivated management with extensive experience in sector of activity
- Negotiated and signed agreements in favorable terms
    - Participation in accordance with objective valuation of the business (preferably a minority interest)
    - Representation in the administrative structure of the company
    - Minority protection clauses included in the shareholder's agreement
    - Previously agreed and accepted divestiture plan for both the management team and the remaining shareholders
- Bullnet Capital's ability to add value to the project
    - Opportunity to share the internal expertise of Bullnet Capital with portfolio company management
    - Capacity to influence strategic decisions of portfolio companies
    - Ability to contribute and exploit synergy with other portfolio companies

21.3 Portfolio companies

Since going into action, Bullnet Capital has invested in five companies all in very early stages of development when investment was made:

NetSpira Networks was a developer of software tools and applications for the management of added-value services in the data networks of mobile telephone operators. Based in Madrid, it had offices in Barcelona and London. NetSpira Networks was sold to Ericsson in June 2005, becoming one of the few examples up to now in Spain of a technological start-up backed by a VC fund that was successfully sold to a multinational company, a leader in its sector. The return on investment for Bullnet Capital was close to three digits of IRR.

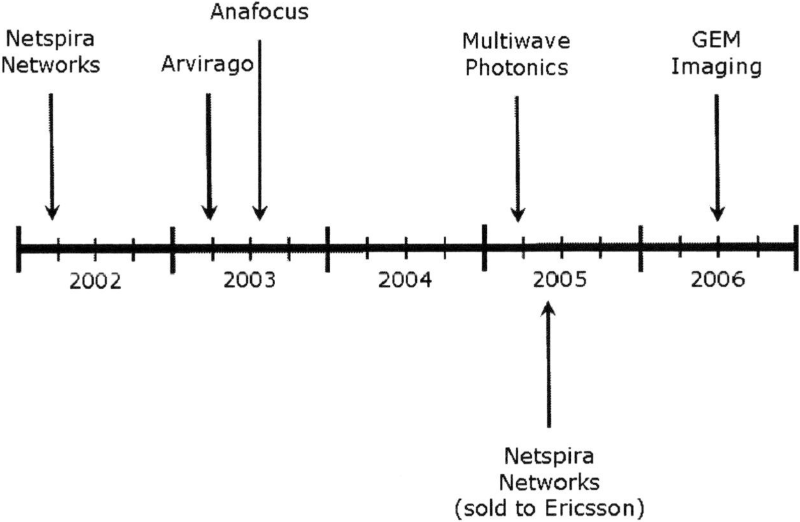

Figure 21.1 Investment/exit dates

Arvirago is a PC and console video game producer. The company is developing a new game called *The Lord of the Creatures*, using its own technology with the aim of achieving outstanding graphic realism with innovative game dynamics. It is based in Madrid and is opening an office in Barcelona.

Anafocus specializes in the design of electronic circuits. In particular, it has proprietary technology for the development of integrated circuits, which allows for the very high-speed acquisition and processing of images with numerous applications in mass industrial markets. It is based in Seville.

Multiwave develops products and systems based on fiber-optic technologies, focusing on two main markets: fiber-optic lasers for scientific and industrial markets and sensor systems based on fiber-optic technology. It is based in Porto.

General Equipment for Medical Imaging specializes in the development, manufacturing, and marketing of nuclear image sensors and equipment for use in medicine and medical/biological research. Obtaining functional images in the areas of oncology, neurology, and traumatology are the main medical applications. It is based in Valencia.

21.4 Key Factors from the Investors' Perspective: Lessons Learned

From our experience over the last five years in identifying, analyzing, negotiating, investing, and following the operations of young innovative companies, there are many lessons learned that are common to all the companies we have looked at or invested in, independent of their particular business model or their sector/industry. These can be classified along the different phases of the life cycle of one of these companies.

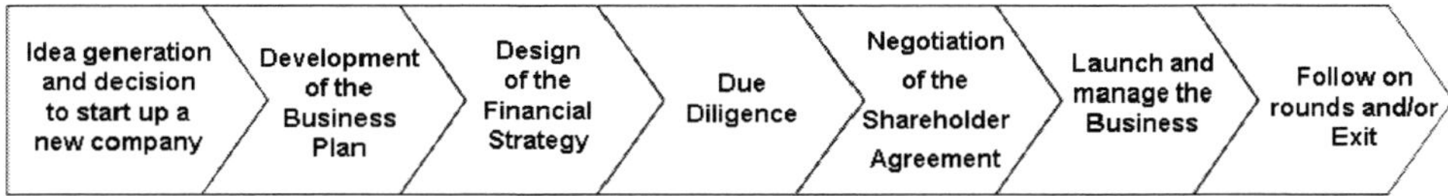

Figure 21.2 Phases during the life cycle of a young, innovative company (from an investor perspective).

21.4.1 Idea Generation and the Decision to Start a New Company

It is very important that entrepreneurs have both their personal and professional expectations quite clear before embarking on the adventure of starting up a new company. Know whether the business project is compatible with said personal and professional objectives. The risk of realizing that you are on the wrong road in life once the company is operative must be avoided at all costs.

It goes without saying that starting is not an easy task, and the journey is normally long and full of obstacles. For this very reason, the entrepreneur should demonstrate a real sacrificial spirit that will overcome those numerous difficult periods that need to be confronted over the life cycle of the company.

One important aspect that the investors value highly is the extent to which the entrepreneur is committed both personally and economically to the project. The more the entrepreneur has to lose if the project fails, the better from the investors' point of view, for the temptation to abandon the project is smaller during the numerous crisis situations that need to be overcome.

Lastly, starting a new company, innovative or not, is a complex process that requires knowledge in very diverse areas. As a general rule, entrepreneurs of this type are usually specialists in a specific technological area but lack commercial or management experience. It is essential that the entrepreneur is naturally disposed to surround himself with complementary individuals and is able to share the responsibilities.

21.4.2 Development of the Business Plan

The business plan is a very useful tool for many things at the same time: to synthesize the basic aspects of the entrepreneur's project, to estimate the financial requirements, to act as an external communication medium, to serve as a reference to which the evolution of the company can be compared, and so forth. It is essential that a good business plan is prepared. To do this, you need to dedicate the time and resources to achieve this. Any shortcut made at this phase in the project will be paid for sooner or later.

The importance of the business plan does not imply that it has to be turned into a philosophical essay. The more concise and focused it is, the easier to read, and by default there is less risk that an investor will reject it after reading the first few pages.

From a qualitative perspective, the business plan should demonstrate that the founding team has the knowledge and experience of the activity sector and the opportunity being presented. It is essential to emphasize the key aspects of the business model. From a quantitative perspective, the base hypotheses must be supported by rigorous analysis, and it is very important to be prudent with forecasts.

21.4.3 Design of the Financial Strategy

One of the main outputs of the business plan must be a rigorous estimation of financing needs over time, without which it is impossible to prepare the appropriate financing strategy.

Financial sources are numerous and varied: oneself; friends and family; business angels; personal credits, subsidies, and grants; venture capital; loans guaranteed with company assets; etc. So as not to waste your (or anyone else's) time, take into account how these financial sources fit in with the state of development of the company. At the same time, know beforehand the expectations of the different fund providers as well as the implication of working with one or another.

Lastly, one aspect that is always important is the dilution (reduction in the percentage size of one's shareholding through the issue of new capital). Normally, this is the most critical aspect from the point of view of an entrepreneur when contemplating the entry of a third party in the capital. But it must not become the only criterion when deciding on a financial source. It must be assumed that sooner or later, the founding team usually loses shareholder control of "their" company while remaining the "engine" of the same (and benefiting from future success).

Taking the philosophy of Bullnet Capital as an example, to us the founder team is key and needs to be motivated by default. It is a critical factor to the success of the investment. For this reason, we try to avoid majority interest in companies in which we invest, at least not in the first round.

21.4.4 Due Diligence

Due diligence is the process whereby an investor investigates the attractiveness of an opportunity, assesses the quality of the management team, and tries to identify the key risks associated with an opportunity.

The more professional the investor, the more sophisticated, long, and complex the process. On the other hand, due diligence has to be thought of by the entrepreneurs as a bidirectional process, for when they are being analyzed, they can (and must) analyze the advantages and disadvantages of having a particular investor as future partners. The main objective of due diligence is the mutual knowledge of both parties.

Bullnet Capital tries to convey to the entrepreneurs that the due-diligence process is not an audit with no added value for them and the company. Rather, it is the opposite; our philosophy is of using due diligence to, among other things, review and contemplate the important aspects of the business model, identify and contact possible clients or commercial partners, attract key individuals who can be incorporated into the team, and synthesize everything into a revised business plan.

21.4.5 Negotiation of the Shareholder Agreement

The norm is that the founder teams do not have previous experience in negotiating the standard terms of a partners' agreement. This generally implies that when the clauses of one of these agreements are read for the first time, there is an urge to reject the investor who, not happy with acquiring a large part of the ownership of the company, demands a series of rights that imply a series of obligations that on first sight are interpreted as abusive by the entrepreneurs. With this in mind, it is important to clarify some points:

- The majority of the clauses included in a partners' agreement are quite standard in the sector, and the differences between one investor and another are minimal.
- If the investor is in the minority (normally the case with Bullnet Capital), one needs to be certain that if there is disagreement with the decisions directly affecting the basic management of the company or corporate aspects, he can exercise the right to veto (with the sole objective of protecting the investment).
- In young innovative companies, investment success depends 100% on the founder team, and therefore it is logical that the investor needs to assure and have the right to request from the team members dedication and permanent commitments, normally guaranteed with shares that they own in the company.
- Venture capitalists are "deinvestors" rather than investors. That is, we have a "temporary" permanent vocation in portfolio companies that could be incompatible with the long-term vocation of the founder team.

And, being in the minority, it is important that there are mechanisms that enable us to liquidate our participation in the case where no agreement is reached with the founder team (tag-and-drag-along rights).

We always recommend that the entrepreneurs look for legal assessors with experience in these types of operations. Not because we want to be up against individuals who negotiate hard, but rather so that they can see that the demands we include in the partners' agreement are quite normal to this operation type.

For all intents and purposes, our experience is that the shareholder agreement is a must but not sufficient to protect our rights as investors. Therefore, for many obligations and commitments, we demand that if there is no consensus with the founder team when taking essential decisions, the agreement is worthless.

21.4.6 Launch and Manage the Business

This is without doubt the most important, longest, and most difficult of all. Trying to be concise without entering into details (there are hundreds of books that talk about the "dos and don'ts" when starting and managing innovative companies), it is important to emphasize the following points:

- The requirements, the environment, and company priorities are very different in each phase: launch, development, consolidation. It is important to know how to adapt the management style, and also that the founder team assumes the fact that it could be important to pass the ball at some point to an outsider.
- It is essential to grow in a controlled manner and rigorously manage the cash-flow. It is paramount that you anticipate possible cash-flow problems in time. To achieve this you have to control expenses and investments—but don't stop (so as not to impede growth).
- The primary error committed by entrepreneurs of this type of company is the tendency to try to achieve the best technological product, in most cases without even knowing what future clients think of it. It is very important to have a commercial mentality from the first day: go to the market.
- As previously mentioned, the consensus philosophy is the most appropriate when entrepreneurs have to coexist with investors. To facilitate this consensus, you must share important decisions with collaborating partners and assessors.
- For small organizations, everyone can add value; that is, you have to know how to delegate and give them space.
- The key to the survival of innovative companies is to appreciate that the primary asset of these companies is the team. Special attention needs to be paid to the selection, motivation, and follow-up of responsibilities.

- We are talking about innovative companies. If you want to sustain a competitive edge, you cannot stop being innovative, and you have to be sufficiently flexible to be able to adapt to an ever-changing environment.

21.4.7 Follow-on Rounds and/or Exit

We have seen that the sourcing of funds is long and complex and requires significant dedication from the management team. The net result, without doubt, is more accentuated once the company has begun to operate and is in a more advanced phase of development. To this end, and as previously mentioned, it is important to anticipate additional fund requirements in time.

If the top-up fund is going to be made by an investor already participating in the company, there are various items to be taken into account:

- From a positive point of view, and thanks to the fact that the investor knows the business, this facilitates and accelerates the due-diligence process.
- There are a series of potential problems that need to be minimized:
    - When negotiating the new value of the business, some conflict could arise between the founders and investors that could have a negative effect on their relationship. To this end, and in order to avoid friction that affects the future of the company, both parties must be flexible and realistic when estimating the value of the business.
    - If the additional funds imply a change of share control, the best part of the partners' agreement will need to be renegotiated because the investors no longer need the minority protection clauses and do not have to concede the founders (new minority) the same rights enjoyed up to now. If these changes culminate in a negative impact on the motivation of the team, it is recommended that alternative legal formulas are found that recognize the appropriate return for the investors for the risk of this new investment, and that the founders maintain apparent control and keep their motivation intact so that they continue to dedicate body and soul to the project.

If the new funds are sourced from a new investor, the situation changes with respect to the previous scenario. In basic terms:

- The due-diligence process will be more extensive and complex than with the original investor in the initial investment. Therefore, be prepared for the possible need of sharing information at some point with third parties (not only with new investors but partners, strategic clients, banks, or subsidiary providers can require information).

- The negotiation process will be much more complex because there will be at least three parties involved: founders, original investors, and new investors. Depending on what item is being negotiated, the interests of one or the other might, or might not, agree with the other parties. To facilitate the process, the founders and original investors must align expectations and objectives.

In the case of a sale (total or partial) to a third party, there are a couple of things that need to be added to the aforementioned:

- It is very important that the activity of the company is not paralyzed, and avoid affecting the team as much as possible until the sale is irreversible.
- Be aware that a possible buyer might request a set of commitments from the management team (business, permanence, etc.) and that the total payment of the sale value will be conditioned by the fulfillment of said commitments.

Acknowledgement

The figures and title in this paper are copyrighted by Javier Ulecia and/or Bullnet Capital and used with permission.

22

Intellectual Property in High-Tech Entrepreneurship

Joseph E. Gortych
Opticus IP Law
Sarasota, Florida, United States

Intellectual property issues are increasingly at the forefront of starting and operating a high-tech business. Views and opinions about the role of intellectual property (IP) in high-tech businesses vary, but it is clear that the role of IP in high technology has evolved greatly in the last decade. Once exclusively the domain of lawyers, IP is moving quickly into the business sphere and is being used and managed by non-lawyer business owners. Because the old-school IP practices are quickly becoming outmoded, high-tech entrepreneurs need to keep current on the modern views of IP and its evolving role in starting and running high-tech businesses.

22.1 IP Views and Opinions as IP Mythology

There seem to be as many different views and opinions about the role of IP in high-tech businesses as there are high-tech businesses. This diversity of opinions is perhaps a direct result of the variety of personalities involved in funding and launching new ventures. At one end of the opinion spectrum is the view of IP as a talisman destined to bring good luck and fortune if vigorously (if not strategically) pursued during the initial phases of a business and beyond. At the other extreme is the view of IP as a black hole into which one pours time and money for no tangible return, to be eschewed not only in the nascent stages of a business, but forever. Most businesses lie somewhere along the continuum between these extremes.

Why the different views and opinions? Some high-tech entrepreneurs have gone through the IP School of Hard Knocks and are now jaded about acquiring IP rights because they do not believe such rights can be fairly enforced. Others have been on the winning end of a licensing negotiation or lawsuit and are now IP swashbucklers. Others have merely heard stories about what IP is, how it is used, and what it does for a company. The different views and opinions passed along over the years by both learned and unlearned elders constitute an IP mythology that includes a body of guided and misguided conventional wisdom.

Making sense of IP reality and how IP actually applies to their particular business situation in the face of the IP mythology is among the many challenges faced by new high-tech entrepreneurs.

22.2 Risk-Seeking Entrepreneurs vs. Risk-Averse Lawyers

At some point in their activities, a high-tech entrepreneur leaves the comfort of his or her linear high-tech world and enters the intertwined and non-Euclidean legal and business worlds. IP is just one of the many different bodies of law—others being corporate, commercial, employment, and export control, to name a few—that constitute a legal n-space that high-tech business people have to deal with in one form or another. The journey into and through this legal n-space is best made in the company of a native guide—an attorney, or a gaggle of the same, often called a "law firm."

There is a natural tension between the risk-seeking entrepreneur and the risk-averse lawyer that starts to become apparent as they journey down the road and encounter business and legal obstacles. This tension can be a source of frustration because the entrepreneur and lawyer have different fundamental goals. The fundamental goal of the lawyer is to protect the company by avoiding risk, while the fundamental goal of the entrepreneur is to grow the company and take risks. Because these fundamental goals are at odds, the high-tech entrepreneur needs to know how to balance legal risk versus business growth. This is true not only in the IP arena but in the other legal areas as well. Good lawyers and good business people know how to strike the right balance and navigate the proper course through the interconnected business-legal maze.

22.3 The Changing Nature of IP in High-Tech Businesses

How is IP used in high-tech business? Fundamentally, the different forms of IP—namely, patents, trade secrets, trademarks, and copyrights—are business tools that can and should be tuned for an intended purpose according to an IP strategy defined by a company's business plans and the IP space of the technology. Too many start-up companies make the mistake of getting IP first and then trying to figure out how to use it. This is the classical patenting paradigm that is as old as the patent system itself. This approach is passively promoted by law firms and is kept in fashion by their disinterest in whether the patents they dutifully obtain for their clients actually have any business value.

A more contemporary view of IP relies on procuring IP rights—particularly, patent rights to inventions—in a directed and purposeful way based on the business plans of a company, the nature and density of the IP space in which the technology is based, and how the IP is meant to further the business goals. But this view needs to be developed, promoted, and driven by the business at the outset. Most law firms, as native IP guides, will provide guidance through the legal morass with consummate skill but will typically not care why you are making the journey or about its ultimate value. They assume you've done your business homework when you show up with your technical inventions that need legal protection. If you reach the IP mountaintop only to discover you've ascended the wrong mountain, your native guides will point to the map you gave them.

22.4 IP Interconnections

Every business organization has a number of interconnected functions. Figure 22.1 is a schematic diagram that illustrates several key functions/departments of a business—namely, product development, marketing, finance, IP, manufacturing, and human resources—and their interconnections.

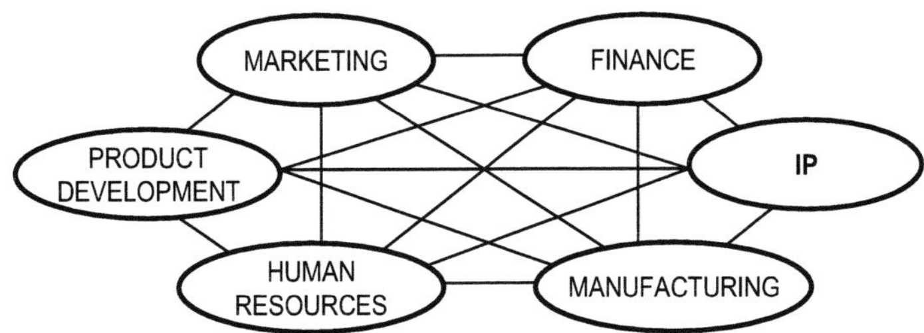

Figure 22.1 Interconnections of the various functions within a company and how they are all tied together and connected to the IP function.

Note that IP is connected to each function. For example, in connection with product development, IP is used to protect inventions embodied in the products as well as to protect improvements to the products. For marketing, IP is used to protect product names using trademarks, to protect written material and designs using copyright, and to brag about patents owned or pending. For finance, IP activities need to be closely budgeted because the amount of money available for IP dictates the IP strategy and goals a business can realistically pursue. For human resources, IP is generated by employees who need training in the basics of IP, who need to sign invention assignment agreements (at least in the U.S.), and whose time to devote to IP needs to be balanced with other activities and responsibilities. In short, IP pervades all aspects of running a business.

22.5 IP Interactions

Every business organization also needs to interact in one way or another with various entities to do business. Figure 22.2 is a schematic diagram illustrating some key IP-related interactions associated with running a high-tech business B.

The IP interaction between business B and its suppliers may involve the need to provide the supplier with confidential information concerning present and future products. In this case, a nondisclosure agreement (NDA) is needed for this interaction to ensure that confidential information remains protected. The IP interaction between business B and its customers may also involve sharing confidential information both ways, which may require a mutual NDA that covers a two-way exchange of confidential information. Customers may also improve upon a business's products and seek to patent the improvements, which can often place business B in an awkward position of competing with its customers. The IP interaction between a business and its competitors can involve IP infringement of the business's IP by the competitor, or vice versa. This interaction can also involve licensing IP (either in or out), including cross-licensing IP to avoid litigation. The IP interaction with the various patent offices of different countries is necessary for a business to obtain patents, trademark and copyright registrations, record patent assignments, etc. The IP interaction with the government involves business regulation, including antitrust issues that can arise when IP is leveraged in a manner that creates monopoly power. The IP interaction involving law firms involves seeking advice and counsel on IP matters and guidance in such activities as obtaining patents from the government, licensing deals, infringement/non-infringement opinions, patent validity/invalidity opinions, etc.. The IP interaction involving industry groups may involve, for example, setting standards for the industry, forming patent pools for licensing IP related to the industry standards, and seeking ways to formulate IP policies that benefit the industry as a whole. The IP interactions related to partners typically involve joint development agreements, dealing with issues of IP ownership for jointly developed IP, how to share and leverage jointly owned IP, etc. In some cases, "partners" may be "customers." The partnership interaction is often a valuable business tool for entrepreneurs to draft off the momentum and size of a larger company and gain access to valuable IP in exchange for providing other valuable IP and/or services.

The processes and procedures a business uses to manage its IP interconnections and interactions constitute IP management. Without a functional IP management program, a high-tech business invites unnecessary risk and introduces tremendous inefficiency in dealing with its IP matters.

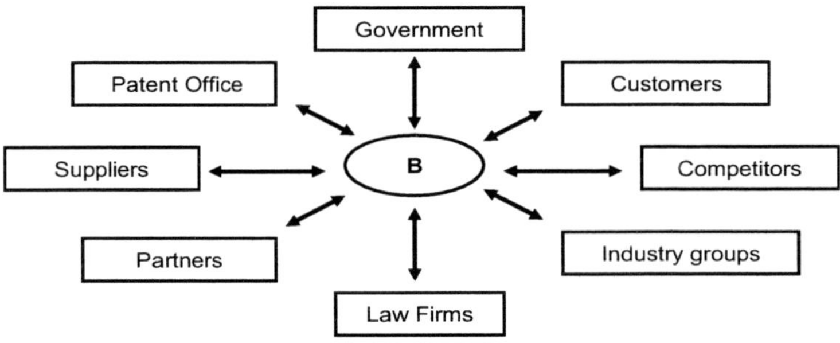

Figure 22.2 Diagram illustrating the key IP-related interactions associated with running a high-tech business (B).

22.6 The Importance of IP Management

22.6.1 The Best-Practice IP Management System

The word "entrepreneur" conjures up a person who is on the cutting edge, is open to new ideas, and who wants to use the most modern approaches and devices to achieve success. A modern cutting-edge approach to IP in high-tech business involves instituting an effective IP management system driven by an IP strategy. A surprising number of high-tech businesses lack such a system and rely on outdated and ineffective ad hoc IP management practices.

For example, businesses often rely on the "bubble-up" approach to managing innovation, in which management waits for technical staff to inform them that they invented something new, rather than regularly and proactively working with the technical staff to extract the company's intellectual assets. Further, high-tech businesses with large patent portfolios are often under the illusion that the mere size of their robust portfolios will protect their freedom to operate (i.e., to make and sell their products) as well as shield them from IP lawsuits, when in fact they are heavily exposed to litigation risk because the patents in their portfolio do not adequately cover their products and services. Furthermore, from a financial perspective, the failure to capture the full IP potential of a business leads to lost licensing and marketing opportunities. In short, an ad hoc IP management approach is reactive and lacks the structure and systematic procedures necessary to compete with more sophisticated players and is seldom successful in maximizing IP value.

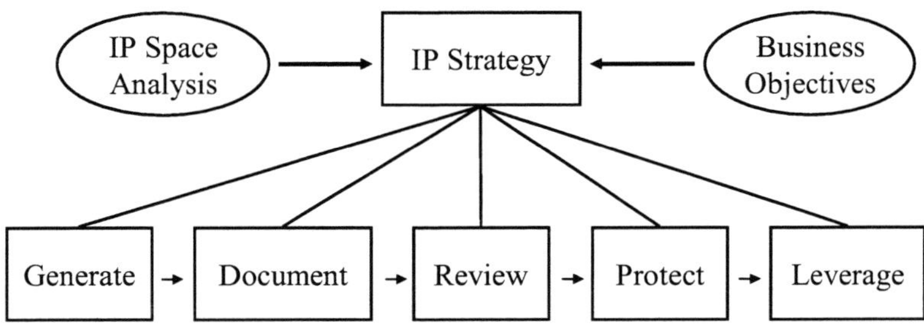

Figure 22.3 A canonical best-practice IP management process.

Yet an ad hoc IP management system is the default system adopted by many entrepreneurs for their start-up companies. Usually, the ad hoc system is then kept as the business grows because there never seems to be enough time to get the IP part of the picture right.

Figure 22.3 is a schematic diagram of a canonical best-practice IP management system.[1] Though not new, the system is apparently not particularly well known or understood by many high-tech businesses, particularly smaller ones and new ones. Although its basic elements are easy to describe in the abstract, it takes genuine effort, dedication, and determination to effectively implement them in the face of other business forces, such as budgets, politics, status-quo inertia, lack of IP education, and insufficient resources.

The best-practice IP management system includes five key functions: invention generation, invention documentation, invention review, invention protection, and invention leveraging. Invention generation is the function wherein inventors (i.e., engineers, scientists, and technicians for technical inventions, as well as salespeople, marketing personnel, and management for business process inventions) create new ideas and inventions in the course of their work. Invention documentation is the function of recording these innovations, usually as "invention disclosures," which can be written by the inventor or drawn up by an IP liaison who interviews the inventor.[2] The invention disclosures are then reviewed, usually by a review board composed of business, legal, and technical personnel. The review board then decides on the best disposition of each invention (i.e., whether it should be patented, kept as a trade secret, sent back to the inventors for further enablement, put on hold, etc.). Some of the inventions are protected (e.g., patent applications are filed for those inventions worthy of patent protection), and those deemed worthy of trade-secret protection are logged and the relevant secret information kept in a manner that will preserve their trade-secret status.[3] The protected inventions are then leveraged in any number of ways. For example, some patented inventions may be leveraged defensively to ensure the company has freedom to practice the invention by preventing others from patenting the same invention. Likewise, certain patented inventions can be leveraged offensively by seeking licenses from others or by simply preventing others from using the invention (e.g., via infringement lawsuits, if necessary). Patents are also collected as trading chips for cross-licensing rather than having to take (and thus pay for) straight licenses from others.

22.6.2 The Role of the IP Strategy

Like any other type of manufacturing process, each step of the IP management process needs to be driven and connected in a sensible way to the other steps. This is accomplished by the IP strategy. Note from Figure 22.1 that without an IP management system, an IP strategy is utterly useless because the IP management system is needed to execute and implement the IP strategy.

The IP strategy is formulated based on two key ingredients: an IP space analysis of the relevant technology and the company's business objectives. It is truly amazing how few high-tech companies—particularly U.S. companies—are willing to spend the time and effort needed to truly understand the IP space in which they do business. Without such knowledge, it is difficult, if not impossible, to formulate a sensible IP strategy for the business. Furthermore, without a sensible IP strategy, attempting to implement a best-practice IP management process—which is key in a competitive high-tech environment—is a daunting task that is usually met with frustration and failure.

One conspicuous sign of a business without an IP strategy is that its patents, when scrutinized, tend to be weak and show signs that they would likely be held invalid if challenged for any one of a number of reasons. They also tend not to be part of a coherent portfolio and are not aligned with the business objectives of the company. Sophisticated IP companies have a strategy to deal with such patents and also take note of who is generating them.

A company's business objectives need to be considered in formulating an IP strategy. Some businesses and organizations expect to generate revenue from IP licensing, while others simply want the freedom to operate. An IP strategy needs to account for how IP is to be used in the business.

The role of the IP strategy is to drive and direct each of the key functions of the IP management process. For example, in guiding the invention-generation function, the IP strategy steers the direction of would-be inventors by telling them the areas in which they are to invent. Specifically, an aspect of an IP strategy might include patenting around the patents of a certain competitor. However, if would-be inventors are not informed of the strategy, they cannot focus their efforts accordingly. This sounds axiomatic. However, the typical high-tech company has no articulated IP strategy—and if it does, it is rarely shared with the technical staff in a way that allows for directed innovation.

The IP strategy also drives the invention-review function to ensure that the inventions aligned with the business objectives are given priority. Here, inventions that are interesting but easily designed around or not relevant to a product to be offered can be filtered out. The protection and leveraging functions are also driven by the IP strategy. In a case where the strategy involves obtaining patents, it is up to the business to ensure that the patent claims relate directly to the IP strategy so as to strengthen the leveraging function—for example, to protect a product that is on the market, or to capture a competitor's product so that it infringes the patent. Many high-tech businesses rely on outside law firms to draft, file, and prosecute patent applications. However, too few businesses—particularly those without in-house patent counsel—take a strong role in managing the direction and content of these applications to ensure that each one is drafted in a manner consistent with the business's IP strategy. To achieve the proper scope of protection, it is critical to share the IP space analysis with the patent draftspersons or to direct them in a manner based on the analysis. This process "tunes" the patents to reinforce the leveraging function, whether it be defensive (e.g., to secure the freedom of action to make and sell a product) or offensive (e.g., to generate licensing income, etc.).

An IP strategy is thus crucial for any high-tech business and particularly so for a new business. A well-crafted IP strategy informed by a proper IP space analysis and an understanding of the business objectives provides six valuable benefits: protection of the company's IP, cost reductions, improved market position, revenue generation, risk reduction, and increased valuation.

22.6.3 IP Sophistication as Business Jiu-Jitsu

One of the benefits of having a thought-out IP strategy and a well-run IP management system is that it imposes discipline on the business and leads to a higher level of IP sophistication. IP sophistication is respected by IP-savvy companies and can be used as an advantage when dealing with less-savvy companies. This sophistication represents a type of business jiu-jitsu that can go a long way to counter size and money differences. Developing a high level of IP sophistication early on in a business is a great way to save money and time in the long run by avoiding common IP pitfalls and taking proper advantage of IP-related business opportunities as they arise.

22.7 Conventional Wisdom with Regard to IP Licensing

Entrepreneurs at some point encounter situations in which IP licensing is an option. For example, an entrepreneur may find that taking a license to a patent portfolio is a less-expensive option than trying to develop related (and non-infringing) technology. In other cases, the IP licensing may be to avoid litigation. In still other cases, the IP licensing may be to grant licenses to others based on IP owned by the entrepreneur.

Technology is typically licensed as patents, though technology licenses for non-patented inventions and know-how are also options these days. The conventional wisdom about patents is that about one-third are invalid for one reason or another and that (conservatively) only about 5% of patents provide real business value to the owner. This combination of poor historical return on investment in patents and the relatively high probability of invalidity provide the context for a would-be licensor or licensee of patented technology.

Anyone who has studied patents and patent file histories in detail knows that patents are like diamonds. While a patent may look good on the surface, a close inspection by an expert can reveal flaws that can be fatal when pressure is applied. The only scrutiny the vast majority of patents receive is that of the patent office examination, which ranges widely in rigor. Many companies rely on the patent office to set the standard for the quality of their patents rather than setting a higher standard themselves through proper IP management practices.[4]

Now, it is true that the minimum government standards for patenting as applied by the U.S. Patent and Trademark Office (USPTO) give each issued patent a seal of approval in the form of a legal presumption of validity. However, it is equally true that the USPTO itself or another governmental body—namely, the federal court system—can revoke the seal of approval of a patent if one can provide enough evidence to rebut the legal presumption of validity.[5] Because the minimum government standards for issuing patents aren't applied consistently, the USPTO seal of approval on many patents is dubious. Said differently, there is a huge problem with the quality of issued patents, the consequence of which is that the ability of a patent to withstand serious legal scrutiny varies widely. This is an important fact for entrepreneurs to keep in mind if they one day find themselves in the position of being either a licensor or a licensee.

22.8 Patent Utility vs. Marketability

Under U.S. patent law and the patent laws of every other industrialized country, an invention is required to be both "new" and "useful."[6] The "useful" requirement (called the "utility" requirement in the U.S.) is intended to prevent the patenting of perpetual motion machines and other non-functional or inoperable inventions. Almost every apparatus that can function or method that produces something real would satisfy the "useful" requirement. However, this utility requirement should never be misunderstood as a marketability requirement. A random perusal of patented inventions will quickly convince one that marketability is most decidedly not a requirement for patenting.

Patented inventions represent the full spectrum of marketability, from the rare but greatly desirable "bottleneck" patents that everyone in the field must use, down to the all-too-common, utterly inane patents that indicate the money would have been better spent on lottery tickets, fast cars, or modern art. The patent marketplace is a veritable Turkish Bazaar with caveat emptor being the order of the day. The abovementioned rule of thumb that only about 5% of all patents provide some real business value means that the patent marketability spectrum is heavily skewed toward the unmarketable. From the licensor's and licensee's point of view, marketability translates into business value (i.e., the value the IP provides to either party).

22.9 Licensee Due Diligence

Given the variability in patent quality and the tendency of patents to lack serious business value, it is critical that an entrepreneur, as a potential licensee, perform due diligence on the patent(s) being offered in order to assess both their quality and business value as they relate to the objectives for which a license is being considered. Likewise, it is important that entrepreneurs, as potential licensors, understand their own IP to the same degree so that they can effectively negotiate licensing terms.

Due diligence is a legal expression that, in the context of licensing, describes the prudent investigation of the details behind a given transaction to ensure there are no serious issues that can adversely impact the deal. IP due diligence in the context of patent licensing includes the following four main activities: establishing IP ownership; reviewing the prior art (particularly that not cited in the patent); analyzing the file history of the patent and the scope of the claims; and looking for other shortcomings in the long list of legal requirements a patent must satisfy to be valid and enforceable (e.g., enablement, best mode, written description, inventorship, and inequitable conduct, to name just a few). Another important due diligence activity is searching for patents that are closely related to or even entangled with the patent(s) to be licensed. Patent entanglement occurs when the claims of different patents overlap to such a point that practicing one patented invention results in infringing another. Closely related patents might not overlap but may nevertheless contribute to the formation of a patent thicket, which occurs when a product or service in a technology field can end up infringing several patents even if the claims of the patent are not entangled because the patent coverage is so dense (e.g., covers numerous aspects of a single product).

22.10 Other Considerations for Entrepreneurs

22.10.1 IP Rights in China

Entrepreneurs involved in manufacturing-intensive products often seek to have the products manufactured in the People's Republic of China because of inexpensive labor costs and a relatively talented pool of workers. However, companies in China have a reputation for not respecting the IP rights of others, which can be a big problem if the manufactured products require the disclosure of IP in the form of trade secrets, manufacturing know-how, and patentable inventions.

Of course, there is also the compelling reason to have IP protection in China because its economy has been growing and continues to grow at an impressive rate and because it represents the largest body of potential consumers in the world—consumers that apparently have a healthy appetite for Western products, including high-tech machines and gadgets.

It is axiomatic that if a product or an invention proves to have significant value, others will try to unlawfully copy and sell it, particularly in places such as China that are still in transition from a culture of IP piracy to IP rights enforcement.

Those in the United States who decry such copying conveniently tend to forget that the United States, as a growing country in the 18th and 19th centuries, blatantly stole key technologies from other countries. One notable example of U.S. IP thievery relates to a prominent New England textile entrepreneur who committed to memory the trade-secret designs of English textile machinery prior to emigrating to the U.S. Needless to say, such designs were highly valuable and protected by the textile companies. Émigrés from England were scrutinized for documents and information relating to such technology by the English authorities prior to their departure because they knew very well the value of keeping the technology in England. At the time, textile machines were the high-tech equivalent of today's semiconductor manufacturing tools in terms of their economic importance. The English textile machinery technology magically appeared in the U.S. colonies shortly thereafter and was used not only to avoid having to import textiles from England but more generally helped fuel the industrialization of the U.S. colonies.[7]

Having patent protection for high-technology products is one way a company can try to mitigate the damage from copying and infringement. This is not an easy or affordable thing to do even in industrialized modern countries that have a long history of respecting IP rights. In countries like China, still making the IP transition, it is even more difficult. Thus, the attractiveness of an inexpensive Chinese labor force and/or a huge potential market for products needs to be balanced with the uncertainties in obtaining satisfactory IP rights protection.

On the positive side of the equation, the trend in China is towards increasing respect for IP rights. In recent years, the Chinese government has taken significant steps to reform its IP rights laws to more closely conform to those of other countries. For example, China is now a signatory to most of the major international IP rights conventions, including perhaps (most notably) the Patent Cooperation Treaty. Other steps have been taken as well, including attempts to streamline IP rights enforcement through the use of special IP rights tribunals.

Still, one needs to appreciate that the Chinese legal system is decidedly non-Western. There is no case law system, and the Western principle of stare decisis is not applied, at least formally.[8] Rather, a particular judge interprets the statutes as he or she sees fit to render a decision.[9] Often, the statutes lack the details of how to actually implement the law, leading to quite a bit of latitude in how a particular decision can be reached.

Given that utility patent protection of 20 years is available in China, one could argue that a great deal of progress will be made in China's IP rights development over this time, thereby increasing the value of Chinese patents that are issued today. In the interim, many will be watching how the IP rights of others are handled by the Chinese courts. It is likely that a number of significant IP-related court decisions will be made by the Chinese courts in the coming years, and that these decisions will provide important insight as to the speed and direction of the evolving IP rights scene in China.

22.10.2 IP Litigation

IP litigation is a growth industry. A potential nightmare for every high-tech entrepreneur is being involved in an expensive lawsuit that drags on for years and that sucks money, resources, and enthusiasm out of a company. No sensible high-tech entrepreneur just starting out wants to wind up anywhere near an IP lawsuit; yet, the chances of a high-tech company being involved in IP litigation is relatively high because many high-tech IP spaces are already quite crowded. When IP litigation happens at the start-up phase, it can kill the business before it gets off the ground. The good news for start-up companies is that the likelihood of being sued for IP infringement is relatively small because the activities of a start-up company tend to be below the radar of others, and there isn't enough money at stake to matter.

The bad news is that once there is enough money at stake to matter, and when business activities rise to the level that they start to have an impact on the market share of others, IP owners are quicker to take notice and, if necessary, take action. The free pass that comes with being a bit player in the market expires when money, market share, and exposure reach a critical mass. Said differently, in some industries (such as the semiconductor industry), there is some noise level of infringement that takes place. As long as one stays submerged in the IP noise, the risk of a lawsuit is low. As soon as one's IP infringement rises above the noise, the risk of being tagged for infringement by another greatly increases.

The best way for entrepreneurs to face the prospect of IP litigation is to take a proactive approach to minimizing the risk of being hit with a lawsuit. This can be accomplished via the IP management system discussed above. Specifically, knowing the IP space in which one resides is a key aspect of avoiding patent infringement.

This is particularly true when starting a new business. Let us say, for example, that one believed that the technology called chemical mechanical planarization (CMP) used in semiconductor device manufacturing had a promising future. Let us further assume that one wished to participate in this future and make money from it by starting a CMP-related business. Let us also say that one wished to be involved in selling the polishing pads for CMP tools because these pads tend to wear out and need to be replaced often.

Prior to starting such a business, it might be worth looking to see if anyone has patents relating to CMP polishing pads. A quick patent database search would show that Applied Materials, IBM, Rohm-Haas, and others have essentially cornered the IP market on polishing pad technology. A few well-placed phone calls would confirm this fact. Without a profoundly new CMP polishing pad technology that would be considered "disruptive," forming a CMP polishing pad company based on conventional technology would eventually result in getting hammered by one of the companies that own the IP in this area. In the vernacular of IP, there is very little "green space" left in the CMP polishing pad technology.

There are a number of technology areas whose IP space is equally crowded and daunting, such as the holography IP space, the nanotechnology IP space, the MEMS IP space, and the quantum cryptography IP space—just to name a few. Entrepreneurs need to be smart about IP for the high-tech companies they start. Not having a good map of the IP landscape is a rookie mistake that can quickly lead to irate investors.

What about entrepreneurs enforcing their IP rights against others? If one owns a piece of real estate and others constantly trespass, allowing the trespass will ultimately result in an easement that allows others to continue to use the property without charge. IP rights are like real property rights in that they need to be enforced against trespassers in order to remain effective. This is part of playing the IP game.

While it is not a happy thought to have to enforce IP rights, the task is made easier if the underlying IP is solid. This is yet another reason that it is worth implementing a well-tuned IP management system. When a sophisticated alleged infringer is confronted, that firm will use all the tricks of the trade to undercut the IP to avoid having to take a license or face a lawsuit. If the IP is solid and the infringement blatant, then filing an IP lawsuit (at least in the U.S.) is not an unreasonable option if the accused party refuses to discuss licensing terms or to terminate infringing activity.

On the other hand, weak IP, which is often the result of an ad hoc IP management processes, is much harder to enforce against even a blatant infringer because the weaknesses introduce uncertainties in the litigation process that a sophisticated infringer will know how to exploit to the maximum extent.

In short, avoiding an IP lawsuit involves IP risk management handled via an IP management system. While the risk of being sued can never be reduced to zero, it can certainly be minimized by having a sensible IP strategy that includes dealing with potential IP litigation issues in a proactive manner, such as by knowing the IP space of the technology and navigating it wisely rather than blindly. Likewise, the strategic use of IP lawsuits to enforce IP rights in situations of blatant and substantial infringement by another is necessary to keep the IP rights enforced and respected, and it is part of doing business in high tech.

22.10.3 The Evolving Role of University Tech Transfer

Almost every industrialized country has a system of technology transfer associated with its universities that seeks to foster commercialization of academic-generated technology. In the U.S., the Bayh-Dole Act plays an important role in tech transfer by establishing conditions under which government-funded research can be owned by the research institution, along with conditions under which inventors receive a portion of the financial gain from commercialization.[10] Other countries have similar laws. Even China has fairly developed rules and regulations regarding IP ownership and technology transfer in connection with their ubiquitous government-sponsored research.[11]

The growing emphasis on tech transfer at universities is based in part on the realization that universities can profit enormously from work that in the Jurassic days of IP was given away. Faculty members and graduate students—that is, the generators of the technology—share in the profits generated by the commercialization and so are motivated to develop technologies that are commercially relevant. Tech-transfer programs have fashioned a new breed of "academic entrepreneur"—one that is interested in both university research and technology commercialization. Some academic entrepreneurs draw the line at licensing their technology, while others go further and get heavily involved in starting and running companies based on the developed technology.

It is important to note that, unlike most high-tech businesses, an academic institution rarely is interested in seeking to cross-license. This is interesting because many large companies that operate in densely packed IP spaces simply rely on the size of their patent portfolios to avoid having to pay for licenses by simply cross-licensing to a would-be licensor.

For example, IBM, Intel, and Applied Materials are all semiconductor manufacturing companies that have huge IP portfolios that allow them to avoid paying IP-related royalties. Now enters the university, with its valuable semiconductor IP that one of these types of companies needs—or is already using and thus infringing upon. Alas, the cross-licensing option no longer exists. Accordingly, if the IP is sufficiently relevant, a potential infringement issue cannot be made to go away through cross-licensing. This offers the university an advantage over industry players when dealing with larger companies that otherwise rely on the cross-licensing option to avoid paying royalties.

While a few U.S. university tech-transfer offices are known for being very effective (e.g., Stanford University, to name one), most other offices in the U.S. and elsewhere are still learning the art of how to be efficient and profitable. In the author's opinion, many university tech-transfer offices have not established the proper IP management system and IP strategies that need to drive their tech-transfer efforts. As a consequence, many university tech-transfer programs follow the old-school approach of "ready-fire!-aim." That is, they come up with inventions ("ready"), then patent everything that comes their way ("fire!"), and then later try to figure out what to do with the patents ("aim").

The better approach is to strategically develop commercially relevant portfolios designed to optimize licensing potential by knowing what to invent. Here, academic entrepreneurs can help by insisting that the patenting process be part of a strategically managed IP system that optimizes the chance that someone would want to license the technology. Anyone familiar with academic institutions will understand, however, that ego-driven politics can be particularly acute and undermine efforts to institute the needed business-like discipline in the tech-transfer office.

22.11 Summary and Conclusions

IP is part of today's high-technology business, and entrepreneurs need to understand and appreciate its changing role. Once essentially the exclusive domain of lawyers, IP is increasingly being managed and leveraged by business people who understand and appreciate its evolving role. The emerging role of IP as a sophisticated business tool having a variety of aspects presents a challenge to high-tech entrepreneurs equal to the technical challenges they face. Entrepreneurs bear the burden of establishing and running productive IP management systems for their businesses that drive the IP generation, documentation, review, protection, and leverage functions. They also have the burden of developing and implementing an IP strategy to drive their IP management system, including development of sufficient knowledge of the IP space of their technology. Strategic IP management optimizes the IP connections and IP interconnections associated with every high-tech business. IP licensing is de rigueur in high-tech business and requires strategic thinking combined with a healthy dose of due diligence. Procuring IP rights in emerging markets and economies such as China presents both opportunities and challenges that require balancing risks and rewards. IP litigation is part of doing business in high technology, and the risk of being at the wrong end of an IP lawsuit needs to be actively managed. On the flip side, the need to initiate an IP lawsuit against infringers may be unavoidable when trespassers become uncooperative and obtrusive. Academic entrepreneurs are enjoying increasing success as university tech-transfer offices expand their role in commercializing academic-generated technology in pursuit of profits. Academic tech-transfer activities represent another facet of the rapidly evolving nature of IP in high-tech business.

Notes and References

1. The notion of connecting the IP strategy to all parts of the IP process was first introduced to the author by John Cronin, Managing Director of IP Capital Group, Inc., Williston, Vermont. See http://www.ipcg.com/whatyoucando/Maximize_IP.Value.htm for more comprehensive views of the connection between IP strategy and the IP management process.
2. IP liaisons are used extensively in Japanese companies, and their use is becoming more prevalent in U.S. companies.
3. Generally, this involves keeping the information secret using reasonable efforts, such as limiting access to the information to those with a need to know, exchanging the information with third parties only under a nondisclosure agreement, etc. See Perritt, H. H., *Trade Secrets—A Practitioner's Guide*, Practicing Law Institute, New York (1994).
4. Gortych, J.E., and Abilock, H., "Managing innovation in the holography business," *SPIE Proc. 6136*, 613605 (2006).
5. The USPTO can also revoke its own "seal of approval" through a process called reexamination under 35 U.S.C. §302 et. seq.
6. 35 USC §101
7. Perritt, H. H., *Trade Secrets—A Practitioner's Guide*, Practicing Law Institute, New York, pp. 16-17 (1994).
8. Latin phrase that represents the legal doctrine of adhering to or otherwise abiding by decisions made in past related cases.
9. Chong, E., "Patent right enforcement in China," *Les Nouvelles*, Vol. XLI, No. 2, pp. 94-102 (June 2006).
10. 35 U.S.C. § 200 through § 212
11. See, e.g., Ngan, A., "From conception to commercialization—university technology transfer practices in China (including Hong Kong), Japan, Korea and Singapore," *Les Nouvelles*, Vol. XLI, No. 2, pp. 68-71 (June 2006).

23

Support for a Young Company

Chris Gracie
Chief Executive
Scottish Optoelectronics Association
United Kingdom

This chapter describes how Scotland has successfully developed a small but vibrant research base to establish a valuable niche in the global optoelectonics market.

23.1 Introduction

Every new company must start with a person or small group of people with an idea. Every budding entrepreneur feels that he or she alone appreciates the opportunity. A colleague to share the idea and vision is ideal support, and family and friends can be vital in maintaining the entrepreneur's drive. However, support in the community should not be overlooked because it can mentor, advise, and provide valuable assistance.

This chapter is based on the experience and activities of the Scottish Optoelectronics Association (SOA). It describes how the association supports its member base and how it interacts with other groups and agencies to further economic growth. Associations in other parts of the world will operate differently, and the support they provide will not necessarily replicate that described here. However, knowing how SOA supports the optoelectronics community in Scotland may allow the reader to ask directed questions that may result in identifying similar mechanisms that they may access.

The Scottish Optoelectronics Association is a membership organization open to companies, research groupings, and individuals who share a desire to promote optoelectronics in Scotland. Although the geographic confines of Scotland identify its membership, SOA is outward looking and has established links to similar associations in the rest of the UK, throughout Europe, and the rest of the world. These links have been put in place in order to further the basic aim of the SOA, which is to support its members and further the growth of Scotland's optoelectronics activity. This aim will be explored in this chapter, but first it should be noted that SOA operates in close liaison with other bodies, and it is important that companies understand who does what. It is also important that each body understand its role to avoid duplication and confusion in the community.

Scotland is split into two regions, each with a government agency responsible for economic development, which for convenience will be called a Local Development Agency (LDA). SOA works closely with these agencies and indeed receives a limited amount of funding from them in order to perform certain tasks. However, the LDA has to consider the strength of Scotland in all technology and business areas; hence, SOA must ensure its member interests remain a high priority with the LDA. SOA's role with respect to the LDA is therefore to provide an input on policy, a conduit to industry, information on industry trends, identification of technology convergence by working with other industry groups, and the identification of interventions, investment areas, and other opportunities that the LDA might pursue. Efficient liaison is achieved by the permanent appointment of an LDA representative on the SOA governing council.

SOA also has close links with associations acting in other technology areas and operating in Scotland.

23.2 SOA's Activities: Networking

The basic function of SOA is to bring the community together to enable interaction at all levels. However, some initiative must be made to establish that a community exists. The fact that a community exists should persuade a new company that, as it grows, the location will provide support. In the case of Scotland, the existence of a number of defense companies using laser ranging techniques, and the fact that Scottish universities secured more than 30% of UK research grants in optoelectronics, prompted the LDA and a number of industrialists and academics to get together to explore the extent of the industry. As a result, the SOA was established in 1994. At that time, optoelectronics was a relatively new industry. Although the laser was invented in 1960, and products combining photons and electrons can be traced even before that date, a starting date for the optoelectronics industry could be argued to be 1980, when the Japanese Optoelectronics Industry & Technology Development Association (OITDA) was established. Japan, from that date, built up a strong industry sector that still accounts for about 40% of global sales. At this time, although America and other countries were heavily engaged in research and electronics companies and had already diversified from electronics products to include optoelectronics, it was not until the mid-1990s that the new technology gained recognition. In 1996, the countries with optoelectronics associations (Japan, U.S., Taiwan, and Scotland) met to form what is now the International Coalition of Optoelectronics Industry Associations (ICOIA).

When the SOA was formed, the initial estimates of economic value and numbers employed were vastly underestimated, as numerous areas of activity were not recognized. One of the first tasks of the association was to draw up an extensive database of the companies and research being conducted. This revealed the true extent of activity. It also enabled the areas of interest to be identified, and meetings designed to appeal to the community were arranged. Over the years, SOA has continued to hold meetings where members can meet and exchange views and information. These networking meetings provide an ideal opportunity for the association to share information and to extract members' needs. They also form a solid basis for all SOA activities.

23.3 Logical Support for Young Businesses

In the mid-1990s, Scotland identified a need to balance the activities of its foreign direct investment (FDI) high-technology industry, which was predominantly engaged in manufacturing and the activities of companies that were integrating design and development of products with their manufacture. It was decided that this could be achieved by attracting a different type of FDI that would include R&D departments in the implant. Another obvious way to achieve this was to transfer to industry the abundant technology produced by the research at the universities. A structured support system has, therefore, been built up to assist the seamless transfer of applicable research to market. It is recognized that the UK research councils perform an excellent function in financing "blue sky" research, and that this is supplemented by an infrastructure funded by the Scottish Higher Education Funding Council. In addition, European research grants are available in key areas. There is, therefore, a wealth of research available for commercialization. This, guided by the overall direction from UK Foresight and European Priorities adapted to Scottish strengths, forms the focus that keeps the research broadly relevant to the future needs of society.

The coincidence of the founding of the SOA and the drive to commercialize research in optoelectronics has meant that the support schemes that have evolved have been rapidly adopted (if not pioneered) by the optoelectronics community. The most recent initiative has been the formation of three intermediary technology institutes with the purpose of funding market-driven pre-competitive research chosen from the blue sky research base. It is planned that this will result in products brought to market between three to seven years in the future. The market areas in which they operate are energy, life sciences, and tech media—all areas where products can be enabled by optoelectronics. The intellectual property produced is offered to industry in order to grow the Scottish economy.

Closer to market is the proof-of-concept initiative. This scheme funds a researcher to prove the feasibility of a concept approximately two years from market with the aim of transferring the technology to industry through either a start-up company to exploit the concept or a license arrangement to an existing company. To further assist the researcher to develop the skills required in producing a business plan for an embryonic business, there is the enterprise fellowship scheme that pays a researcher to take a sabbatical year to learn commercial skills. More recently, this concept has been widened to include industrial fellowships allowing those employed in industry to evolve a business plan based on a new idea or non-core technology within their company. Although this is a relatively new scheme, it is anticipated that because the individual has industrial experience, a new company that results from the scheme may develop more quickly then one from an academic background.

Another significant feature, especially obvious in the optoelectronics area, has been the establishment of technology transfer departments by many of the universities. These departments work closely with industry in developing research and exist at the Universities of Strathclyde, Glasgow, St. Andrews, Abertay Dundee, Dundee, Paisley, and Heriot-Watt. Recently, cross-disciplinary departments have been formed (e.g., cosmic at the University of Edinburgh). The breadth of technologies covered makes them ideal for addressing the overlap of technologies or "white space" opportunities. In the case of cosmic, areas such as biophotonics can be addressed by the team comprised of experts in biotechnology, material science, and nanotechnology.

SOA's own contribution to technology transfer is embodied in the Technology Transfer in Optoelectronics and Microelectonics (TTOM). This scheme is based on networking meetings, bringing industry and academia together with the objective of identifying for companies the technology needed to take a new product to market and matching the need to an academic provider. SOA can seed-fund the feasibility study to prove that a solution has been identified. The company can then embark on developing the product with internal or investment funding.

These initiatives have assisted in the impressive record of an average of five optoelectronics start-up companies each year in Scotland.

23.4 Information

An important role of any association is to be an information source for its members. Enquiries can frequently be answered through industry knowledge, but when an answer is required in some depth, or is in unfamiliar territory, it is important that the association refer a member to a likely source that can supply an answer. An association needs to build up a database of information covering the research at universities, the funding mechanisms applicable, training courses available, sources of market intelligence, companies on the supply chain, and contacts throughout the world. This entails a great deal of work, but reference to the database of knowledge may mean an inquiry is answered in a few minutes.

Another important activity that requires input from members is the annual survey. SOA conducts both an industry and academic survey to define regional activity and its economic impact. Although this requires members to devote a little time to form filling, it does give them a benefit in return that may not be, at first sight, obvious. The information compiled shows government agencies, investors, and both suppliers and buyers the strength of the industry. This in turn generates confidence in investors and provides evidence of the long-term stability buyers and suppliers require. Government policy is influenced by the community's capability of delivering economic growth, hence investment from this source is also guaranteed. The annual survey demonstrates the depth of expertise and workforce, which emphasizes the stability of a region. Members benefit from the credibility arising from the availability and circulation of this information worldwide.

23.5 Marketing

Marketing a company's products is vital, and the success of this department will dictate the growth rate of a young company. Previously, emphasis has been placed on the transfer of technology from the research base. This implies a technology drive to the development of products. However, market pull will decide whether a product remains in small-scale manufacture or advances to volume production and secures a significant market share.

Even before a company is started, the potential market must be identified, but a new company that spins out of a university will probably concentrate initially on product development and the technology embodied in the product. The key personnel at this stage are in technology. However, as the product development nears completion, the key posts transfer to manufacture and marketing, hence it is not unusual to see changes in management at this stage. This is not necessarily a bad move, as most people perform best at a particular discipline. They are then free to move on to develop a new product or start up another new company, becoming serial entrepreneurs.

SOA has participated in a number of schemes to help its members identify and penetrate markets. Market trend and analysis information is gathered from all SOA's sources and made available to members. In addition, SOA identifies exhibitions and conferences at which a significant number of its members can benefit from attendance. In collaboration with the LDA and Scottish Executive, a presence is established at relevant events and companies are assisted to attend. The assistance varies from funding obtained through government agencies for new, small firms to introductions for the established larger concerns.

SOA considers the establishment of a showcase of UK companies and research to be important, so it collaborates with other bodies to stage exhibitions and conferences in the UK. This activity falls into two areas: attracting international conferences and exhibitions to Scotland. Recent examples are

- The 17th Indium Phosphide and Related Materials Conference, an international conference sponsored by the Institute of Electrical & Electronics Engineers (IEEE), the Lasers and Electro-Optics Society (LEOS), the Institution of Electrical Engineers (IEE), the Institute of Physics, and the Electron Devices Society and held in Glasgow in May 2005.

- The 17th International Conference on Laser Spectroscopy held in Cairngorms National Park in June 2005.
- EuroDisplay 2005, the European Exhibition and Conference of SID, held in Edinburgh in September 2005.
- The European Conference of Optical Communications 2005, held in Glasgow.

The second part of this activity is to take an active role in the organization of the biennial conference and exhibition, the Photon Series. Photon02, 04, and 06 were held in Cardiff, Glasgow, and Manchester, respectively.

To complement the above, an initiative of the LDA has been to set up a global Scot network throughout the world. The network is of expatriate Scots who have been successful abroad and are willing to help Scottish companies with market advice and introductions in their adopted country.

A new venture for SOA is the exposure of its members to mentoring in product management and market channel management. As stated, most new technology companies have a technology bias to their management. This new venture is to introduce the marketing aspect at an early stage.

23.6 Roadmaps and Market Overviews

Technology planning is important for many reasons. Globally, companies face many competitive problems. Technology roadmapping, a form of technology planning, can help deal with this increasingly competitive environment. Looking ahead, a company should be able to estimate a product's life cycle and identify what R&D is necessary to launch new and successor products.

A company might have its own roadmap; probably most large companies do. However, small companies might not have the time or resources to produce their own. Hence, an association can produce a roadmap for its members. SOA's experience has been that the roadmap development process can inspire member companies to initiate their own internal roadmapping process. It has been found that roadmapping can bring together both individuals and companies to combine expertise in producing a new product.

Another benefit from roadmapping is that the supply chain has been strengthened because a component manufacturer targeting one market and a systems integrator in another market have been brought together to their mutual benefit.

A final benefit is that gaps in infrastructure have been identified, and government support was secured to fill those gaps.

The roadmap produced by an association follows the same structure as that produced by an individual company, but it is designed to influence a broader audience. The roadmap considers what the market need will be in 10 or more years. This is not an easy task and can, of course, be completely wrong; an example is the telecommunications crash of 2000. The future may be postulated through the evolution of technology or by scenarios. Once products that meet market needs are defined, the time required to develop the product, identify the components, suggest the necessary research, and establish the infrastructure can all be shown on a timeline. This timeline can then be used by companies all along the supply chain to plan their product development. It can be used by universities to align their research and by government agencies to invest in support mechanisms.

Networking meetings can form the basis of discussions between the different users of the roadmap to test its validity and check that individual actions will be backed by supporting actions from others.

23.7 Establishing Infrastructure

One output from a roadmap can be the identification of gaps in the local infrastructure. In the years leading up to the telecommunications crash in 2000, a newly formed company was able to attract ample funds to invest in the capital equipment required by their business. However, since 2000, investment levels in new companies have been much lower, hence capital equipment is more difficult to finance for individual companies. The value of networking in this case is significant to a business: a number of young companies, all needing similar capital equipment, can present a collective case whereby the government can, by itself or acting with private interests, provide facilities for them to prototype their products. Examples of this in Scotland are facilities for silicon microelectronic, III-IV compound semiconductor, and packaging prototyping. Although primarily for Scottish companies, these facilities are available to companies all over the world and allow a young company to postpone large capital investment until sales justify the expenditure.

23.8 Cross-Disciplinary Assistance

Optoelectronics is an enabling technology, and products embodying optoelectronics can be found in ever-increasing market areas. Optoelectronic products therefore more often than not rely on other areas of technology. For this purpose, SOA has joined with other associations operating in Scotland to run joint meetings, thus extending the networking for its companies and enabling cross-sector partnering. Examples of this are meetings held with the microelectronics, wireless, medical device, and software associations. Resulting from these meetings are frequent requests for TTOM funding where, for instance, a medical device company requires optical expertise to develop or improve a product to gain competitive advantage.

23.9 Skills and Training

The education delivered in the UK is comprehensive but cannot respond to new requirements if these are not known. As a relatively new industry employing fewer people than established industries such as banking, optoelectronics must make its training needs known and arrange for a critical mass of students to ensure the economic continuation of training. SOA joins with other technology bodies to influence the take up of mathematics and science at school and university and encourages the establishment of specialist courses. At university level, SOA has been involved in setting up a master's degree in display technology. This degree is obtained by securing the necessary number of modules in different display technologies, each offered by the UK university recognized as the expert in the particular technology. The teaching modules are supplemented by a project undertaken at a displays company. SOA is currently helping to spread the course throughout Europe by increasing the modules available and involving European universities. This should ensure that the student numbers are maintained at a viable level to meet the needs of a growing European displays industry. In addition, SOA promotes specialist PhD and engineering doctorates in relevant subjects.

A university degree is not the only level of skill required by the industry. Associations must predict members' needs in other areas and attempt to develop the necessary training. A sister association has recently developed a comprehensive training system for fiber-optics operatives involving training providers at all levels. As optoelectronics grows, similar courses will become necessary. An example of the result of not planning ahead occurred at the peak of the telecommunications bubble in 2000, when no available operators could handle small-diameter fiber. Neither were there any training courses to convert electronics operators. In such a situation, a single company cannot predict the total local demand, and this responsibility falls on the association. This information can be obtained from the skills survey data held by the association.

23.10 Promotion and Lobbying

Most companies are aware of the advantages of the press release, which brings them to the attention of potential employees, investors, collaborators, customers, and suppliers. It falls to the association to promote its members as an economic force, adding to the individual public relations a critical mass that attracts wider attention. For this reason, SOA provides articles for the trade press and gives interviews to the media. In addition, SOA contributes to the local news magazine *Scottish Technology News* (STN). STN gathers the relevant PR from companies and adds industry overviews.

The association undertakes a lobbying role locally in Scotland, nationally in the UK, and within Europe. The objective of lobbying is threefold: to retain optoelectronics in the forefront of policymakers' minds, to ensure Scottish R&D

features in UK and European programs, and to attract investment into Scotland. The first objective has been examined previously. The second involves taking part in the formulation of photonics strategies for the UK and Europe. These strategies give rise to strategic research agendas that shape the calls for R&D proposals. SOA then acts to alert members to the calls for proposals and assists in the formation of collaborative teams to bid into the programs. The third objective is advanced as SOA speaks to potential investors to attract them to Scotland. In this context, investors can either be interested in investing in individual companies or be companies interested in locating in Scotland. In the latter case, after initial interest has been generated, the LDA and the Scottish Executive are involved.

23.11 Product Development

Young companies may not only need access to capital equipment but also to product assessment facilities. SOA has therefore entered into a European partnership that will buy pre-competitive components from young companies and give them to selected research establishments to assess. The project's objective is to put pre-competitive photonic components and systems in the hands of researchers and students, at no net cost to the university or to the company that furnishes the prototypes. As a result, students are trained on the next generation of emerging technologies and products as identified by European industries. This training orients students towards advanced technology jobs in Europe, thus helping to develop a highly educated and productive workforce.

Each company that participates in the program (especially small to medium enterprises) has a new and valuable resource for implementing research and development at a reduced cost, which is also precisely focused on the products and issues that are most relevant to that company's continued growth and success. The project, named ACCORD, takes its inspiration from similar optoelectronic components and exchange programs that have been demonstrated in Japan and the United States. This activity illustrates the importance of the worldwide networking activities carried out by SOA. The ACCORD project has benefited from extensive discussions with the Optoelectronics Industry Technology Development Association (OITDA) in Japan and the Optoelectronics Industry Development Association (OIDA) in the U.S.

The OITDA managed an exchange program in the mid-1980s called the Joint Optoelectronics Development Project (JODP). This initiative was part of the 6th-Generation Computing Project, funded by the Japanese ministry MITI. The program had an international component, and the National Science Foundation (NSF) and Defense Advanced Research Projects Agency (DARPA) from the United States were invited to participate. U.S. participation was brokered and managed by the OIDA. The initiative, while successful, was terminated in the early 1990s with the end of the 6th-Generation Computing Project. In 2000, the OIDA launched its own components exchange program called the Photonics

Technology Access Program (PTAP). The OIDA was able to benefit from its previous experience with the OITDA to improve both the attractiveness to participants and efficiency with regard to results. The PTAP program is still financed on a seven-year renewable contract from DARPA and the NSF.

23.12 Standards

An association is sometimes the only body with enough resources to develop standards that can help develop and protect its members' businesses. A standard opens a market to a wider buying public, giving opportunity to all in that market. A constant threat is the development of a standard that precludes a company's participating in a market, so each association keeps a watch on standards development and inputs information where its members have an interest. This activity is more prevalent and urgent in Japan than in Europe because a great deal of development is carried out in that region.

23.13 Summary

In this chapter, the activities of SOA in Scotland have been described. These activities have been designed to be part of the overall UK approach to optoelectronics. Associations in other parts of the world may perform differently according to practice in their own countries. However, although the operational detail and scale may be different, most of the activities undertaken by SOA will be repeated elsewhere, depending on the maturity of the industry.

Each year, SOA meets with all the members of the International Coalition of Optoelectronics Industry Associations (ICOIA). At this annual event, each association presents its activities over the previous year and plans for the future.

Each association has gained immense benefit from sharing these plans, and governments, companies, and research agencies have also recognized the economic and market advantages gained from the collective wisdom and knowledge of this global community.

Part IV

The Universities

Two visions, from Europe and the U.S., concerning the role of a modern university in entrepreneurship, based on knowledge and technique, are presented in this section.

Universities are primarily educational institutions. Their major role continues to be as the principal conduit through which bright young people mature into their role in society. Within universities, these bright young people are educated through a discipline and often practice their passions through the student union, the debating society, and sports clubs, which demonstrate organizational, political, and leadership skills along the way.

Universities have over the past half century or thereabouts emerged into another role, essentially that of catalyzing economic development within their local communities. During the early days of Silicon Valley, Stanford University played a central role, predominantly as the intellectual resource. The university's technological influence also rapidly emerged as an important stimulus—all that university research really could be applied. Stanford was among the early pioneers of the university "contracts office." That university's Office of Technical Licensing is now arguably the exemplar to which other institutions aspire.

Much has been written of the complex dynamics of Silicon Valley, which has spawned an infrastructure of technological and managerial talent, adventurous venture capital (much of which is now invested worldwide), and an adventurous local marketplace. Silicon Valley is without doubt a unique phenomenon. Universities worldwide have observed this nucleation process with—let's face it—envy as, indeed, have national and regional politicians and economic development agencies. Consequently, attempts to emulate Silicon Valley have sprung up worldwide, in the process emphasizing that Silicon Valley really is unique.

But as we have seen in this book, academic institutions are the stimulus for a great deal of spin-out activity, providing a vehicle for exporting technology research in university laboratories into useful and economically positive products. A few academics have also gained financially (modestly, in most cases), and their institutions have benefited through a mix of equity investment, licensing opportunities, and, possibly most important of all, simply through recognition of their catalytic role in local economic development. University attitudes towards associated commercial development have switched during the past half century as these benefits have become more evident to university administrations and have also become recognized within the external political community.

The "contracts office" is, however, a very mixed story. There are maybe a couple dozen institutions worldwide whose commercialization activities are really economically and socially beneficial and within which the contracts staff are creative and managerially astute. The comments from the authors of the Mirada case study, for example, specifically mention the role of Isis Innovation in Oxford. For many, though, there remains mutual suspicion, perhaps even antagonism between the contracts department and the academics, with a muddling of roles. Who is really responsible for the corporate interface? Who should talk to the venture capital? How do technology academics find the management skills? How do we really define realistic expectations on the returns on often-modest intellectual property?

Many universities are associated with a local science park developed under the auspices of local and regional government. In themselves, these science parks have nucleated local technical innovation and attracted talented young people. The tenants of these science parks are usually a mix of academic spin-out and other high-technology companies. While accurate statistics are hard to find, the casual survey seems to indicate that the majority populating most science parks have their origins elsewhere. This by no means detracts from their value. It simply indicates that the stimulus provided through the academic institution has a much broader influence than is perhaps immediately obvious.

So the roles of universities in high-technology entrepreneurship are mixed and complex. While the university's short-term objective is often to benefit, for reasons political and economic, from the scientific and technological research within its campuses, the consequences of this endeavor are often far reaching. The structure and management of this process continues to progress along an often-tortuous learning curve. The last couple of contributions to this book present snapshots—one from Charlotte, North Carolina, the other from Madrid, Spain—of this activity from those closely involved in its operation.

It seems very likely that the structure and management of this interaction between the academic community and the high-technology entrepreneur will continue to become increasingly important and will substantially enrich the societies and communities within which it is implemented successfully. It is a highly tangled subject with unique solutions required in individual communities within their particular complex mix of cultural, economic, political, and social environments. It is just one of the many facets that we have attempted to cover in this short text within the complex story of high-technology entrepreneurship.

24

Strategic Support: The Case of the Technical University of Madrid

Gonzalo León Serrano
Professor and Vice President for Research
Polytechnic University of Madrid
Madrid, Spain

The basic missions of a modern university must include and stress the exploitation of R&D. Entrepreneurship using new knowledge and techniques is being promoted by means of several procedures. This chapter briefly addresses a vision of the Spanish situation, focusing on the case of the Polytechnic University of Madrid.

24.1 The Evolution of the Third Role of Spanish Universities

A modern university must combine four basic missions: to teach consolidated knowledge (both in pre- and post-graduate curricula), to generate new knowledge as a by-product of its research activity, to transfer and exploit new and pre-existent knowledge to productive sectors and public administrations, and to disseminate scientific and technical knowledge to society at large. These are not isolated functions, and it is a responsibility of universities to balance them in order to satisfy their social role. The relative importance of each varies widely with the type of university and its relationship with the industrial context.

As Figure 24.1 suggests, enterprises interact with universities in the implementation of all mentioned missions but in different ways. In teaching activities, enterprises may provide funding, organize, or provide experts in courses or seminars adapted to their specific needs. Usually, this cooperation is focused on post-graduate courses, although the involvement in PhD programs is lower than expected.

In knowledge generation, enterprises cooperate by funding contract research or participating in collaborative research projects. This is commonly carried out through bilateral contract-research schemes. The abovementioned cases of research contracted through a company have strong limits in their ability to protect and exploit the university's results. Nevertheless, R&D projects funded by regional, national, or European agencies generate results that can be directly exploited by the university to support technological innovation.

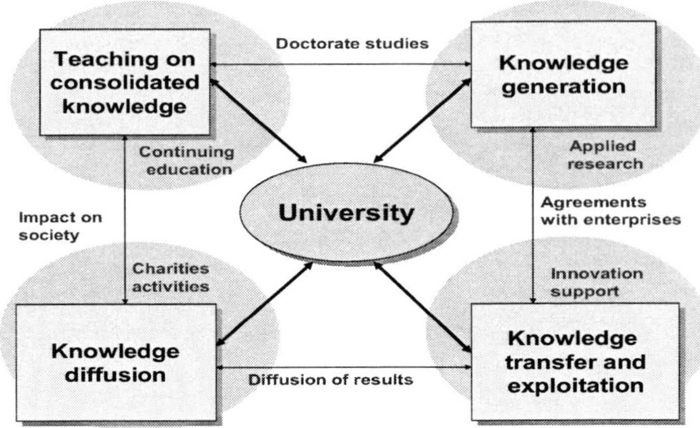

Figure 24.1 University missions and the cooperation space with industry.

Finally, agreements with private foundations or other nonprofit organizations can support the diffusion of results to society. Unfortunately, this last element is not very common in Spain.[1]

Figure 24.2 depicts a more detailed view of the activities framed in the knowledge transfer and exploitation mission: contract-research, commercialization of technology, mobility of researchers to enterprises, and creation of new technology-based firms.

Universities can support all the activities depicted in Figure 24.2 by using their own internal R&D structures (departments, R&D centers, research institutes, etc.) or by creating specialized units. This is the case in the majority of the Spanish universities.

From the historical standpoint, the solution in the 1980s was to create internal offices (termed in Spain as "Offices for Technology Transfer," or OTRI) partly supported by the national and some regional R&D plans. This instrument was a valid one but with some limitations for supporting permanent cooperative structures with the industrial sector. It was easier to know and disseminate the offer provided by the university than to bring industrial demand to the research groups at the university.

The success in the development of these four elements varies widely depending on the type of university, the institutional involvement, and its regional industrial context. Unfortunately, with the exception of bilateral contracts for R&D projects, the other activities had limited development. Licensing is a marginal activity in economic terms (royalty incomes are very low in Spanish universities), and mobility to the industrial sector is more a personal adventure than an institutional priority.

During the last decade, Spanish public administrations have progressively adopted a more active role in this field through the creation of specific policies.

The main goal was to increase the relationship between universities and enterprises and to foster an entrepreneurial role in universities as a new strategy to strengthen the industrial tissue. This movement towards entrepreneurship will be addressed in the next section.[2]

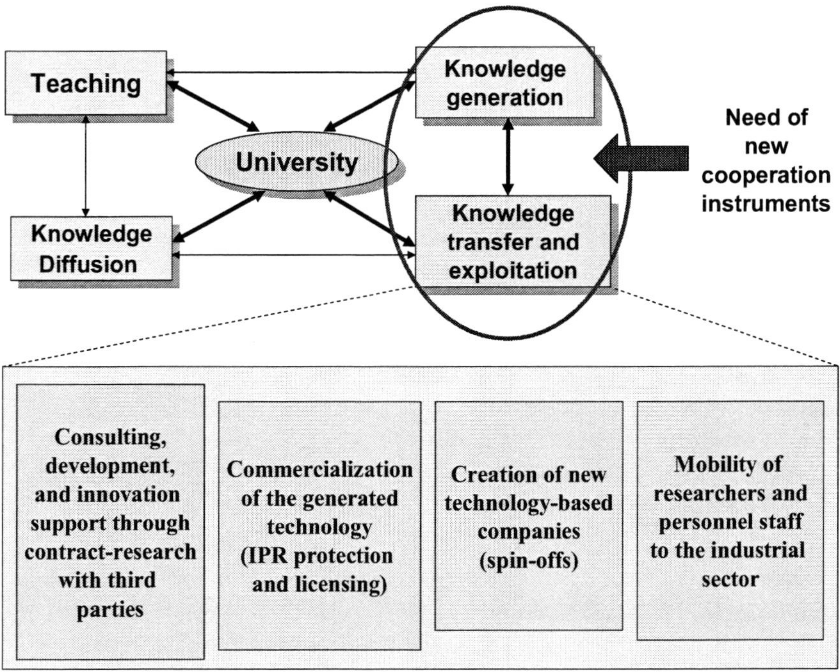

Figure 24.2 Activities related to the third role of universities.

The most relevant instruments for stimulating R&D cooperation in this context were the support for the creation of university-driven scientific and technological parks, and setting up large, consortia-based R&D projects in which universities played an important role (bringing to Spain the successful model used in the EU Framework Program). In all these cases, the objective was focused on evolving the conventional cooperation between universities and enterprises based on short-term R&D project-to-project cooperation plans into one based on the creation of long-term strategic alliances. This evolution was also enforced by enterprises as a part of the implementation of the open innovation concept, once they were convinced that any institution could possess all the required knowledge in-house.

24.2 The Next Step: Entrepreneurship

In the last decade, a new strategic option has been added to the scenario presented above to improve the so-called "third role" of universities.[3,4] Spanish universities recognized that they could better satisfy their social role if they could directly exploit their R&D results by contributing to the regeneration of the industrial tissue. As a consequence, new programs to support the creation of new technology-based companies (NTB spin-offs) were launched with the support of public administrations.

These programs were based on three shared assumptions: it is an institutional responsibility of universities to promote entrepreneurship between students and faculty members; an opportunity exists to support NTB spin-offs in incubators as a part of the university-driven science and technology (S&T) parks under development; and it is possible to attract risk capital to support the early phases of NTB spin-offs created from the R&D activity in the university.

The implementation of these three assumptions has been a moderate catalyst increasing the numbers of spin-off created by Spanish universities. It is also relevant to mention that an indicator related to the number of NTBs spun off from public research centers has been included in the strategic goals of the national R&D plans since 2000. Some support offered to these programs demonstrates the growing involvement of the Spanish government as well as some regional governments in this objective.

Unfortunately, success in NTB spin-off creation not only depends on the availability of specific funding or incubation infrastructures but on an attitude change among all stakeholders. It will require much more time and some regulatory changes in university law and statutes.

Statistically, Spanish students do not consider the creation of their own company a valid mechanism for starting their professional life in comparison with students in other countries. Not surprisingly, it also happens with Spanish PhD students who are mainly oriented towards filling the ranks of faculty members in the public system rather than participating in spin-offs. This is one of the most important reasons for the low numbers of NTB spin-off creation from universities. The same situation occurs with faculty members (civil servants by law) whose potential motivation to be involved in one spin-off did not compensate the risk of abandoning their stable position at the university. This specific case is being addressed in the reform of the University Act under discussion at the Spanish Parliament.

The typical spin-off evolution represented in Figure 24.3 opens the discussion on the area where the university, as a public institution, should participate. From the accumulated experience, the framed area in the figure seems the most convenient for two reasons: the activity of university spin-offs in their first phase is usually based on technology previously created at the university, where later the activity will be based on its own technology developed as a new spin-off company, and the activity of the company can be located in areas controlled by the university (incubators or space in the S&T park) when the link with research groups from which promoters came or the access to facilities is still necessary.

Less agreement exists on the role of a public university in risk venture-capital companies. It is true that one of the weaknesses of the innovation process in Spain is the lack of sufficient seed capital to finance university spin-offs during the first stages, but how to overcome this problem is still open.

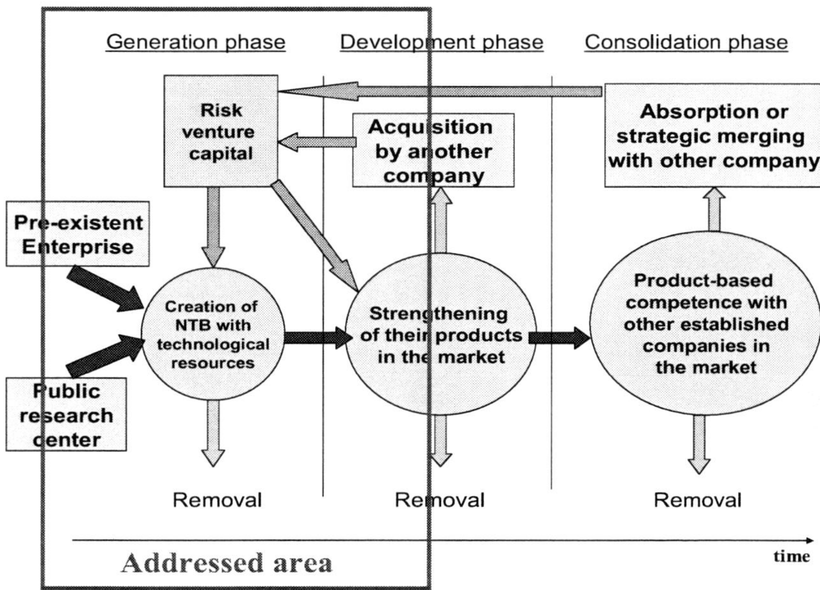

Figure 24.3 Evolution of spin-off support.

In 2005 and 2006, some initiatives were implemented to promote the creation of specific technological funds. In one, several universities (18 in August 2006) have created a company to manage a risk venture-capital fund (UNINVEST). The Spanish government has also created a "fund of funds" to facilitate risk reduction for potential investors.

Anyway, the present situation is that the majority of Spanish universities have included in their strategic plans specific programs to foster cooperation with the industrial sector, the development of scientific and technological parks, the management of spin-off incubators, and, in a limited number of cases, the participation in venture-capital management firms. The combination of all these elements implies a deep cultural change in universities that occurred in a short period of time.

24.3 The Case of the Polytechnic University of Madrid

The Polytechnic University of Madrid (UPM) is the largest technical university in Spain. With 42,000 pre- and post-graduate students, 3,300 faculty members, and 2,500 administrative and support personnel, it has an annual budget of €360 million, where more than €100 million comes from external sources for R&D. It covers all branches of engineering and architecture in three- and five- to six-year curricula in 20 technical schools as well as a faculty for sports activity. UPM is a relatively new university (established in 1971), although some of its engineering schools have more than 200 years of history.

The connection to the industrial sector and to public administration is very strong, and 60% of the external R&D resources are related to contract-research in cooperation with industry. As an example, UPM has the largest number of contracts in the EU FP6, and is the second largest in terms of economic returns. Annually, more than 700 contracts are signed with private companies. Fifty-four chairs were sponsored by private companies in 2006. This funding allowed the recruiting of more than 200 engineers with fixed-term contracts for R&D projects.

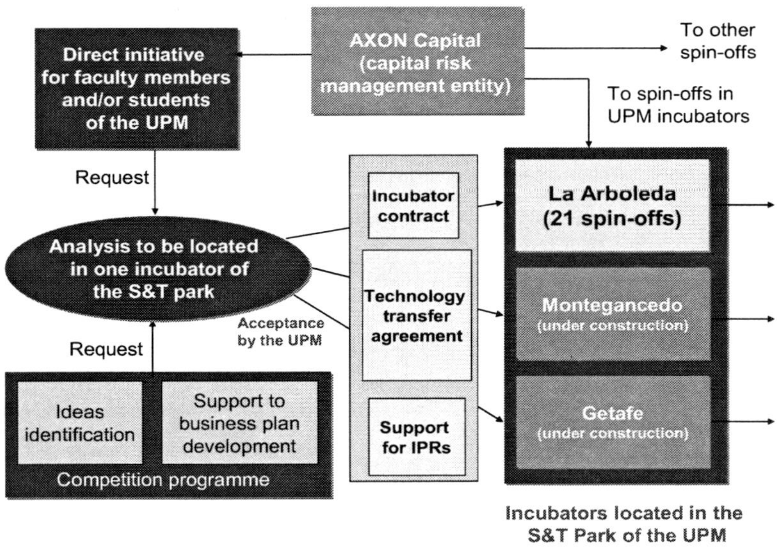

Figure 24.4 NBT creation program at UPM.

This experience and motivation was also the starting point for developing an entrepreneurship mentality. This is not a completely new process; as an example, two companies created years ago now have hundreds of employees (ISOFOTON in the photovoltaic sector and GMV in the space sector). The difference was that these initiatives were bottom-up processes without an institutional program to support them in the initial phases.

This NTB spin-off program was created four years ago, and Figure 24.4 summarizes the activities carried out under the program. The competition program annually promotes the identification of 60 to 80 business ideas from students and professors; some of them (more mature or with more potential to become a new spin-off) are selected for a specific training and tutorial support process for consolidating business plans. The best business ideas receive a prize and have priority in finding a place in the incubator if the spin-off is created (prizes are conditional upon the effective creation of the company). Additionally, not shown in Figure 24.4, entrepreneurship seminars have been prepared for pre-graduate students in all technical schools, and a mentoring system for students has been launched with the cooperation of the regional government of Madrid.

Presently, only one incubator (located in the S&T park) with 21 spin-offs is in operation (more recent spin-offs are in independent locations outside incubators). Two more incubators were to be operational in 2007 and 2008 in different sites of the S&T park, increasing the total capacity to 50 spin-offs.

Notice that if a request to be located is accepted by UPM (not necessarily from the competition program), promoters need to sign a contract to be located in the incubator as well as a technology-transfer agreement (if it was not agreed on before the request) to ensure the correct use by the spin-off of the background technology previously created at UPM.

To facilitate the funding of NTB spin-offs, UPM created a new venture-capital management company, AXON Capital, in July 2006 with other private investors, TECNALIA (a network of technological centers located in the Basque country), and Quantica (a similar company in Italy closely linked to the Italian National Research Council and Italian universities). The objective is to allocate an investment fund of €15 million in 2007 and €35 million in 2008 to fund NTB spin-offs, mainly from the Spanish public system (not only from UPM).

In some cases, UPM participates as a shareholder of the new company; in others, it simply acts as a facilitator by allowing its personnel to be involved. In all cases, the objective is to sell its participation as soon as the spin-off is consolidated. The value of the participation is linked to the value of the technology provided to the spin-off. In all cases, an agreement should be signed to regulate intellectual property rights.

There is a complementary approach that UPM is exploring: the creation of new legal entities (as private firms or foundations) with other external partners. Here, the objective is to be involved permanently by a legal entity in order to exploit some technological knowledge. Two examples of this case are AIDIT S.L. (with the Universidad Politécnica de Catalunya), offering R&D project evaluation services to access the Spanish R&D tax deduction system, and CENTESIL S.A., to develop new ultrapurified silicon for photovoltaic solar cells with ISOFOTON, TECNICAS REUNIDAS, DC WAFERS, and the Universidad Complutense de Madrid.

This effort has been accompanied by the development in the last two years of a set of internal rules and procedures to support the creation of new R&D centers, the management of IPRs, and the institutional support to create and participate in NTB companies.[5,6] More specifically, there is an institutional will to create joint R&D centers with other well-established companies, moving from short-term project-to-project cooperation to long-term strategic alliances. We feel that this scenario will offer new opportunities for the development of S&T professional careers to UPM post-doc students.

This attitude is also a consequence of the high level of research cooperation with the industrial sector and the wide acceptance in all actors in the UPM community that, as a technical university, this is the best way to be useful in Spanish society. The combined effect of the development of the S&T park (with three spin-off incubators), the new internal regulations, and the support from the regional government paved the way for success. UPM is one of the Spanish universities that places greater emphasis on its strategic plan to entrepreneurship—an example of institutional leadership.

24.4 Conclusions

This article has presented the emergence and evolution of entrepreneurship in Spanish universities. The approach used was to demonstrate the relationship between the institutional acceptance of the third role of universities and the emergence of innovative attitudes.

From a historical perspective, based on the experience gained through cooperation with industry, new steps towards entrepreneurship have been addressed by Spanish universities. This challenge could not be implemented without a sufficient understanding of the previous steps and strong institutional support for the exploitation of research results. The role of public administration in implementing new cooperation instruments and specific funds for spin-offs is also very important.

Entrepreneurship requires a rich and diverse landscape to grow. The efforts carried out in recent years by creating NTB spin-off incubators and the emergence of risk capital funds oriented to early stages is creating an adequate scenario to dramatically increase the number of spin-offs in coming years. It is obviously easier in technical universities, or in cases such as information technologies where the physical barriers to create a new company are lower.

The case of UPM, as a technical university, is used as an example of institutional involvement in strengthening cooperation with industry and NTB spin-off creation. The use of the development process of the S&T park as a catalyst for entrepreneurship shows that deep institutional involvement can obtain results in a very short time.

A lot of work is still pending in Spain to facilitate R&D cooperation between public universities and private enterprises. Nevertheless, a plethora of policy instruments are available to enable the transition from cooperation based on individual (and many times short-term) research projects to a long-term, stable cooperation based on common assumptions and interests in strategic alliances. This is a new strategic scenario that universities, companies, and public administrations increasingly support.

I am confident that Spanish universities will assume the necessary institutional leadership to better contribute to the construction of the knowledge society by assuming entrepreneurship as one institutional objective. Then, figures of the new NTB spin-offs and technology licenses will demonstrate clear advances a few years from now.

Acknowledgement

The figures and title in this paper are copyrighted by Gonzalo León Serrano and used with permission.

References

1. Castro, E., Fernández de Lucio, I., "La I+D empresarial y sus relaciones con la investigación pública española," *Radiografía de la investigación pública en España*, Biblioteca Nueva, Madrid, Spain (2006).
2. "Public-private partnerships for research and innovation in Spain: background and issues for discussion," *Working Party on Innovation and Technology Policy* (December 2004).
3. "Implementing the community Lisbon Programme: Fostering entrepreneurial mindsets through education and learning," *Commission of the European Communities*, Brussels, Belgium (2006).
4. "Delivering on the modernization agenda for universities: education, research and innovation," *Commission of the European Communities*, (2006).
5. "Normativa de propiedad intellectual" ("Intellectual property regulation"), *Consejo de Gobierno de la UPM* (December 2005).
6. "Normativa de creación de empresas de base tecnológica" ("Regulation for the creation of technology based enterprises"), *Consejo de Gobierno de la UPM* (April 2006).

25

University Research and the Optics Industry

Faramarz Farahi
Professor and Chairperson
University of North Carolina
Charlotte, North Carolina, United States

The structure and management of the interaction between the academic community and the high-technology entrepreneur will continue to become increasingly important and will substantially enrich the societies and communities within which it is successfully implemented. The research universities have played a large role through their students and faculty in establishing start-up companies. In this chapter, an analysis is presented of the key factors for optimizing the transference results by means of the creation of new photonic companies.

25.1 Introduction

Throughout the world, and particularly in the United States, small businesses play a major role in economic growth. In developed countries, one significant area for small-business development is in science and technology. In this area, research universities have played a large role through their students and faculty in establishing start-up companies. Many examples support this assertion. Spinning off new ventures from research institutions has played a key role in the development of high-technology clusters in areas such as Silicon Valley and Boston. It is also believed that university research affects private-sector innovation. Little is understood about either the specific channels through which these effects occur or their impact among various industries. However, it is obvious that public research affects industrial R&D in a wide variety of manufacturing industries, and publications and conferences are the most important knowledge-flow channels for these effects in most industries.[1]

Before academic research results can be commercialized, the technology or knowledge has to be transferred from the research organization to industrial/commercial entities; this process is called technology transfer. Any managed technology-transfer activity is likely to be both costly and peripheral to the main purpose of the university, which is to develop and disseminate knowledge.[2] This activity also requires a different form of managerial structure and style than is inherent in the institution. Although Rogers et al. have identified five technology-transfer mechanisms from university to industry—spin-offs, licensing to existing companies, cooperative R&D agreements, teaching and publication, and interaction and cooperation—the main focus of this article is on the role of universities in spinning off technology companies in general and optical companies in particular.[3]

25.2 The Current Situation

In recent years, an avenue that has been discussed more often in the business community is the formation of companies that evolve from academic institutions through commercialization of intellectual property and transfer of technology developed within such institutions. However, many of these companies did not result from the structural process of research commercialization. Often, these companies were founded by entrepreneurs despite the research institute with which they were associated. An example of such a company is Digital Optics Corporation (DOC), a spin-off from The University of North Carolina at Charlotte (UNCC). This company was founded in 1990 by a faculty member and his students while the university had little or no internal policy to manage such a technology transfer. Only recently, research institutes have developed proactive policies to stimulate the commercial exploitation of public research, through spin-out formation. At the same time, changes in the institutional environment have made such a policy possible. Despite these changes, detailed knowledge of the processes of proactively spinning out new technology-based companies from research institutions remains in short supply.[2]

In the United States, particularly after the passage of the Bayh-Dole Act in 1980, universities increased their efforts in formal technology transfer and licensing. The Bayh-Dole Act has had a significant impact on the way universities view their role in the commercialization of their technologies.

The act states, "It is the policy and objective of the Congress to use the patent system to promote the utilization of inventions arising from federally supported research or development; to encourage maximum participation of small business firms in federally supported research and development efforts; to promote collaboration between commercial concerns and nonprofit organizations, including universities; to ensure that inventions made by nonprofit organizations and small business firms are used in a manner to promote free competition and enterprise without unduly encumbering future research and discovery; to promote the commercialization and public availability of inventions made in the United States by United States industry and labor; to ensure that the Government obtains sufficient rights in federally supported inventions to meet the needs of the Government and protect the public against nonuse or unreasonable use of inventions; and to minimize the costs of administering policies in this area."

Bayh–Dole dramatically changed incentives for firms and universities to engage in technology transfer. It simplified the technology-transfer process by instituting a uniform patent policy and removing many of the restrictions on licensing. Since the passage of this act, U.S. universities have increased their efforts in formal technology transfer and licensing, and in some cases, investments in new firms. In the first 20 years after the passage of this act, the number of universities engaging in technology licensing has increased eightfold.[1]

Although the Bayh-Dole Act has removed a major barrier to engaging universities in technology transfer, many other factors have an impact on this process. W. B. Gartner lists 10 ingredients that are most important to successful venture creation: suitable financing, availability of a competent workforce, access to helpful suppliers, government support, the proximity of universities to assist in research, the availability of land or facilities, access to transportation, support of local population, available support services, and low entry barriers.[4] Each of these factors is important; some of them may be more critical to one venture than to another, but all of them have a considerable role in ensuring new venture success. Proximity to universities is noteworthy in two respects: in high-tech start-ups where new inventions or technologies play a dominant role, these institutions can make a significant contribution to success through research, problem solving, and engineering support.[5] Another benefit of the university is its business school and the resulting availability of consulting services in the areas of marketing, production systems, accounting, and finance, as well as access to expertise in the area of information technology.

Mokry has added two factors to Gartner's 10.[6] The first is the existence of an entrepreneurial subculture. The tremendous success of Silicon Valley and Boston support this assertion, as it very much supports the notion that entrepreneurs feed off each other in a synergistic fashion and create their own dynamic environment. Mokry's second factor is incubator organizations, many of which are initiated by local universities and governments. Recent studies have shown a more positive awareness of educating people in entrepreneurship and how to become entrepreneurial. It has been determined that potential entrepreneurs can be encouraged through university-based entrepreneurship programs, entrepreneurship within an established definition can be taught, and entrepreneurial alumni do succeed and provide further insights and educational materials for dissemination in the classroom.[5] For university-initiated start-ups, in addition to these factors, the role of the office of technology transfer is critically important. Creating a successful technology-transfer office requires the integration of a number of factors. A successful office would employ qualified personnel, particularly including individuals with good business ideas and a deep appreciation of technology and the challenges that a newly formed company faces. This office should be, in principle, a resource for entrepreneurs rather than a source of income for the institution. In particular, the nature of the optics industry, which is highly concentrated in some areas and very fragmented in others, requires that the technology-transfer office have personnel with knowledge of the field who could facilitate the formation of new start-ups accordingly.

New optics firms, like other new technology-based companies, should exploit radical technologies with broad-scope patents to compete with established enterprises, meaning that new firms founded to exploit university inventions will be more likely to survive if they possess these attributes. Radical technology undermines the advantages that established firms have in making incremental improvements to their technology; it undermines the competence of established firms and turns their existing customer relationships into liabilities rather than assets.[7-10] In addition, the new optics companies should consider the nature of competition from established players in the product market. Such competition will determine whether a firm formed for exploiting radical technology will survive entry because the introduction of radical technology by a new company may spur action by established firms that inhibit the survival of new firms.

To survive, new firms must first build an organization and acquire assets that will be used in conjunction with their radical technology. This process is more difficult in the concentrated sectors than the fragmented sectors of the optics industries. Examples of concentrated sectors of the optics industry are optical fibers for telecommunications and flat-panel displays. One can argue that more fragmented sectors (i.e., optical metrology, optical sensors, and to some extent optical components) are still areas where small firms are very active and there are spaces for new entries. First, in concentrated sectors of the optics industries, the marketing and manufacturing assets necessary to exploit a technology lie in the hands of a few established firms, which tend to acquire ownership of these assets to mitigate contracting problems.[11,12] Because new firms do not have these assets in place at the time of founding, they need to build them. When the needed assets are controlled by a few large, existing companies, the new firms have fewer parties to work with. This increases the difficulty of establishing an agreement with one of them to obtain needed assets.

In contrast, in fragmented sectors of the industry, the assets necessary to exploit the new technology are available from a wide number of industry players, minimizing bargaining problems in efforts to obtain access to these assets.[2] Since the average size of firms is larger in more concentrated segments, the new firms require larger marketing and manufacturing assets.[13] Given the state of the capital market, new companies find it difficult to build up these assets on a scale that is cost effective with established players, undermining their ability to survive.[14] In the concentrated sector, established enterprises have cost advantages and market power, allowing them to drive out new competitors.[15] Finally, in fragmented sectors, new firm entry does not necessarily impact the efforts of market leaders to serve their customers; while in concentrated sectors, the establishment of the new company impacts the efforts of the market leaders because the target customers of the new firms belong to large established players.[16] As a result, entry by a new enterprise in a concentrated sector is more likely to invoke retaliation by a large established firm, and the new company finds it hard to enter the market successfully and its survival is impaired.[17,18]

Various researchers have argued that new technology firms will perform better if they have broad-scope patents because strong intellectual property protection is necessary to protect their technology from imitation while they create the marketing and manufacturing assets necessary to exploit their technologies.[19] This is contingent on the entrepreneur founding a company in a fragmented sector of the industry. In a concentrated sector, the difficulty of creating these assets makes this strategy problematic. Yet, as we indicated above, to survive, new firms must develop manufacturing and marketing assets that are used in conjunction with their new technology. Broad-scope patents facilitate this transition because they provide better protection than narrow-scope patents. As Merges and Nelson explain, "The broader the scope, the larger number of competing products and processes that will infringe the patent."[19]

Optics spin-offs from universities differ from other university spin-offs in two fundamental ways. University research and the resulting technologies typically have limited scopes. This is in contrast to general-purpose technologies in which start-ups have an enhanced chance of survival because these technologies can be used in a variety of applications. The latter provide new firms with flexibility that is useful to overcoming technical and market risk.[20] Most technological innovations in the field of optics with a broad scope of applications are being led by major firms. This is in part due to the nature of the field and in part to the fact that each university with active optics research has focused in a narrow domain. A very good example is Southampton University in the UK, which has an excellent reputation in the field of fiber optics, but it has either none or few activities in areas such as light emitting diodes, micro-optics and optical integration.

Also, optics start-ups tend to require far more capital to commercialize their technologies than most university spin-offs and often require sizable venture investment to put their products on the market. As a result, the survival patterns for optics firms might be systematically different from the survival patterns of other university spin-offs.

Starting a company, as opposed to licensing to an established firm, may be the best course to follow when the tacit knowledge of the inventor is indispensable for development.[21] Start-ups suit some technologies better than others. Certain technologies, useful for many applications, may not be cost effective for any single firm to license; disruptive technologies might be resisted by existing producers. In addition, barriers to entry must be sufficiently low. These conditions seem to prevail in optics, most likely due to technological complexity in general and the difficulty of the development process in some cases.

25.3 Actions to Remove Barriers

The number of new optics start-ups is not limited by the availability of good ideas; rather, it is limited by the inability to overcome barriers that exist between people with good ideas and the resources needed to obtain proof of concept. One should emphasize that universities are already the institutions that develop individuals who can extend their knowledge to recognize opportunities where others don't, individuals who don't think in conventional, linear terms but who excel in thinking at higher levels of complexity. But, there are certain actions that universities, research institutions, and other state and federal agencies can take to help remove these barriers and provide the right climate for the number of optics start-ups to increase while improving their chances of success, hence igniting regional and national economic growth. Some of these actions are:[22]

1. Universities typically earn profits from their inventions through royalties and have policies that divide these profits between the inventors and the university. The distribution of royalty rates between inventors and the university could influence the tendency of entrepreneurs to start a business and exploit university inventions. Hence, universities could be flexible in this regard, based on the need of each particular start-up, to encourage entrepreneurship activities.
2. Start-ups that use incubators have significant cost advantage over those who don't. Incubators allow young firms to develop technologies in close proximity to inventors and the infrastructure needed for such a technology development. Therefore, the use of incubators should increase the optics start-up rate. A good incubator is an organization that accelerates and systematizes the process of creating successful enterprises by providing them with a comprehensive, integrated range of support, including space, business support services, and clustering and networking opportunities. Universities can either lead such incubators or work closely with them by making their extensive resources available to start-ups.
3. A university that takes an equity stake in start-ups in exchange for paying patenting or other up-front costs could facilitate the formation of start-up companies. University equity investments made in lieu of paying patent costs or up-front license fees reduce a company's cash expenditures, facilitating firm formation.[23] This is something that universities should seriously consider.
4. Any mechanism that provides seed funding and, furthermore, provides funding to develop the innovative technology to more advanced stages could make the acquisition of capital easier for optics start-ups. States provide a variety of financial incentives to attract companies within their borders to generate jobs and ignite regional economic growth. The level of funds required to support a new optics firm is minute in comparison with those incentives. Therefore, it is prudent policy to provide funds for promising optics start-ups at their very early stages.

25.4 Conclusion

Good applied research generally requires the ability to quickly determine the key bottlenecks to finding a solution and applying a variety of techniques to overcome these bottlenecks. This can mean applying well-known techniques to a new problem, applying a new technique to an old problem, or even applying a new technique to a new problem, which has the most risk but the greatest potential for changing the way people think.[24] Good policies are more or less the same. In this case, it requires identifying the key barriers facing new optical technology firms and considering a variety of methods to overcome these barriers. This can mean applying well-known or new methods to help with optics start-ups. The latter has the greatest potential for aiding in the start of a successful cluster of optics companies in a region, which will completely change the dynamics of economic growth and job creation in that region. We have seen the way electronics, pharmaceutical, and software industries have redefined the economic parameters of certain regions in the United States. Optics has similar potential, but it requires creative regional policies that will allow the full potential of this untapped technology to be explored.

References

1. Mowery, D.C., Shane, S., "Introduction to the special issue on university entrepreneurship and technology transfer," *Management Science*, Vol. 48, No. 1, p. v–ix (2002).
2. Shah, D., "Success analysis of start-ups in the field of microsystems and manotechnology in the UK," *MS Thesis* (2004).
3. Rogers, E.M., Takegami, S., Yin, J., "Lessons learned about technology transfer," *Technovation* (2001).
4. Gartner, W. B., "What are we talking about when we talk about entrepreneurship?" *Journal of Business Venturing*, Vol. 5, p. 15 (1990).
5. D'Cruz, C., O'Neal, T., "Integration of technology incubator programs with academic entrepreneurship curriculum," *PICMET 03*, p. 327-332 (July 2003).
6. Mokry, B. W., *Entrepreneurship and Public Policy: Can Government Stimulate Business Startups?*, Quorum Books, New York (1988).
7. Nerkar, A., Shane, S., "When Do Startups That Exploit Patented Academic Knowledge Survive?" *International Journal of Industrial Organization, Elsevier*, Vol. 21(9), p. 1391-1410 (November 2003).
8. Utterback, J., *Mastering the Dynamics of Innovation*, Harvard Business School Press, Boston (1994).
9. Christensen, C., Bower, J., "Customer power, strategic investment, and the failure of leading firms," *Strategic Management Journal*, Vol. 17, p. 197 (1996).
10. Tushman, M., Anderson, P., "Technological discontinuities and organizational environments," *Administrative Science Quarterly*, p. 439 (1986).
11. Williamson, O., *The Economic Institutions of Capitalism*, Collier McMillan, New York (1985).

12. Teece, D. "Profiting from technological innovation: Implications for integration, collaboration, licensing and public policy," *Research Policy*, p. 285 (1986).
13. Mansfield, E., "Composition of R&D expenditures: Relationship to size of firm, concentration, and innovative output," *Review of Economics and Statistics*, Vol. 63, p. 610 (1981).
14. Geroski, P., "What do we know about entry?" *International Journal of Industrial Organization*, p. 421 (1995).
15. Kamien, M., Schwartz, N., "Market structure and innovation: A survey," *Journal of Economic Literature*, Vol. 13, p. 1 (1975).
16. Eisenhardt, K., Schoonhoven, K., "Organizational growth: Linking founding team, strategy, environment, and growth among U.S. semiconductor ventures, 1978-1988," *Administrative Science Quarterly*, Vol. 35, p. 504 (1990).
17. Romanelli, E., "Environments and strategies of organization start-up: Effects on early survival," *Administrative Science Quarterly*, p. 369 (1989).
18. Acs, Z., Audretsch, D., "Innovation, market structure, and firm size," *The Review of Economics and Statistics*, Vol. 71, p. 567 (1987).
19. Merges, R., Nelson, R., "On the complex economics of patent scope," *Columbia Law Review*, Vol. 90, p. 839-916 (1990).
20. Shane, S., "Prior knowledge and the discovery of entrepreneurial opportunities," *Organization Science*, Vol. 11, p. 448 (2000).
21. Geiger, R. L., Hallacher, P., Harper, B., Prabhu, R., Sá, C., "Nanotechnology and the states public policy, university research, and economic development in Pennsylvania," *A Report to the National Science Foundation* (July 2005).
22. Di Gregorio, D., Shane, S., "Why do some universities generate more start-ups than others?" *Research Policy* (2001; also in the 2001 Global Entrepreneurship Research Conference).
23. Hsu, D., Bernstein, T., "Managing the university technology licensing process: Findings from case studies" *The Journal of The Association of University Technology Managers*, Vol. 9, p.1 (1997).
24. Winters, J.H., "Reflections on industrial research," *IEEE Signal Processing Magazine*, Vol. 6 (July 2005).

Postscript

Some Concluding Thoughts

Our aim in this book has been to convey the feel for technology entrepreneurship. We decided at the outset to omit the recipes of business plans and presentations, a treatise on legal responsibilities, employment, safety, public liability, professional liability, and all these other essential intrusions into the technology adventure. Our hope, however, is that we have conveyed the impression that this is an interesting, exciting, and viable approach to technology exploitation, that it has many, many facets and no stereotypes, and that, most of all, it is stimulating and enjoyable.

We have also omitted many of the experiences that we could have covered. The infamous optoelectronics boom and bust of the early part of the 21st century was fuelled by misplaced optimism in the information exploration, by totally unrealistic faith in projecting excessive exponential growth into the future, and frankly by straightforward financial greed, particularly from the investor communities. We all have friends and colleagues who were at the receiving end of excessive investments and saw the shop closed down on them within a couple of years or less. We hear, though, much less of the stories of the rise from the ashes. We know well one fiber-optic company taken over by a corporate megalith during the boom, expanded out of all recognition, and closed two years later. Then the core staff engineered a management buyout, and through steady growth, patience, wisdom, and parsimony are now employing a hundred or more and, we hear, attracting buyout interest again. This example is far from solitary and may even be more typical than is generally expected. Indeed, the gloom after the boom that cast a shadow over technology ventures for some time was in practice very much less justifiable than it appeared. Another anecdote on the same theme: in a conversation with the founder and chief executive of another optics company, established long before the boom but seriously threatened in the aftermath, I asked him, "Well, what did you learn from this?"

"It doesn't really matter," he replied. So the survivor survives and it has always been so.

There are many other aspects of technology entrepreneurship that we still must cover. The role of professional societies, exemplified in SPIE, the publisher of this small volume, can be central as a source of information and also—very important—as the organizer of opportunities and venues through which to meet like minds. The standards and regulatory authorities also inevitably play their part.

Most of our enterprises will sooner or later need to succumb to ISO 9000 or its appropriate variant on manufacturing quality. It is far and away better to do this soon rather than later, before the record tracing becomes tedious in the extreme.

We need to take note of the progression of technical standards and also of the social and legal environment as it evolves. At present, for example, the uncertainties but inevitabilities of carbon trading and all it implies will surely offer immense opportunity for the astute technology-oriented entrepreneur. We have also, since the turn of the century, witnessed a social change as security paranoia has gripped whole communities and opened immense new markets. Many of us have our opinions (often less than complimentary) about this, but regardless of the opinion, the fact of the opportunity cannot be denied.

Then there is the extremely important role of partnerships. We have hinted at this, but perhaps it requires more emphasis. You will no doubt discover how it fits through your individual experience, but partnerships through collaborative programs, national and international, and through business collaborations are critical for the small entrepreneurial technology enterprise. There is much to be gained from leverage on others' presence, especially in complementary technologies, markets, and regulatory intricacies.

So, finally, all that we can recommend is that if this avenue feels appropriate, then by all means explore it. It will have its frustrations, it will no doubt offer immense satisfaction, it will be a great way to learn a lot very quickly, it will be hectic—but we do believe that it will be genuinely enjoyable.

The Editors

Appendix

Due Diligence Check List

Mike Morris
Cofounder and President of Ocean Optics Inc.

I. Background Information	
	1. Business plans for the current and next fiscal years.
	2. Industry studies and external reports as well as internal strategic reports, marketing reports, or other product information.
II. Corporate Structure	
	1. List of all jurisdictions where company has offices or distribution relationships. Specify approximate size of office and number of employees at each location.
	2. Organizational chart by legal entity showing country of incorporation and indicating any subsidiaries that are not wholly owned.
	3. Organizational chart by division and department and an organizational chart by name three levels down. Brief biographies of the top-10 employees.
	4. Employee headcount by functional area and engineering, and total turnover rate for the past two years.
	5. Copies of bylaws and certificate of incorporation, as amended to date; minutes from board and subcommittee meetings for the last year; minutes from shareholders' meetings.
III. Financial Information	
	1. Audited financial statements for the last three years.
	2. Accountants' management letters and management's response for the last three years.
	3. Quarterly breakdown of financial performance for the last two fiscal years and the current YTD.

III. Financial Information (cont.)	
	4. Projected income statements, cash flow, balance sheets for the next two years, by quarter, with detailed assumptions.
	5. List of all internal reports utilized by management on a regular basis.
	6. Marketing or operating plans for the current year and the next two fiscal years.
	7. Explanation of revenue and cost recognition methods.
	8. Summary of current bad-debt reserve, recent bad-debt experience, and unusual charges or credits for the last three fiscal years.
	9. Detailed accounts receivable aging schedule for most recent period.
	10. Analysis of cost of sales for last financial year and YTD period.
	11. History of product returns.
	12. List of the top-15 customers over the last three years and any customer over 10% in any given quarter.
	13. Schedule of depreciation and amortization expenses.
	14. Summary of off-balance sheet assets/liabilities.
	15. Details of any contingent liability.
IV. Tax	
	1. All tax returns for the last five years and any prior year still open or subject to appeal or not yet paid.
	2. Any outstanding tax issues and copies of all correspondence relating to outstanding issues.
	3. Details of any dispute with the Internal Revenue Service or any state or local tax authority leading to investigation or appeal.
	4. Details of tax credits or losses available to be carried forward.
	5. Details of any tax indemnities in favor of the company.
V. Marketing and Strategy	
	1. Schedule of proposed product introductions and product-release schedule.
	2. Description of channels of distribution and any significant arrangements.

V. Marketing and Strategy (cont.)	
	3. Internal market size and growth projections and any market-share data.
	4. R&D/engineering priorities and associated R&D budgets.
VI. Shareholder Information	
	1. Summary of authorized capital stock and issued capital stock.
	2A. Current share register.
	2B. Current shareholder account report.
	2C. Stockholder totals listing.
	2D. Active investor list.
	3A. Listing of all share options granted, including any employee options showing number of shares, date of issue, cost, exercise price, and terms.
	3B. Current options register.
	3C. Option holder account summary.
	3D. Option expiration report.
	3E. Board of directors list.
	4. Copy of all equity/option incentive plans.
	5. Copies of any agreements between shareholders and/or employees and company.
VII. Corporate Finance	
	1. Summary of terms of lines of credit, loan agreements, credit agreements, and other debt instruments, including any amendments, renewal letters, notices, waivers, etc.
	2. Other agreements evidencing outstanding loans to or guarantees by the company, including capital lease obligations.
	3. Agreements to acquire or dispose of any assets, subsidiaries, securities, or business, other than in the ordinary course of business.
	4. Detail of investments in other companies.
	5. All material agreements encumbering real or personal property, including mortgages, deeds of trust, and security arrangements.
	6. All documents pertaining to any insolvency proceedings with which the company or any of its subsidiaries or offices or directors have been involved.

VIII. Commercial Contracts	
	1. List of all R&D partnerships and agreements affiliated or sponsored by the company.
	2. List of principal contracts entered into in the operation of the company's business and copies of each such contract.
	3. Forms of supply and manufacturing agreements; key supply and manufacturing agreements; list of key suppliers not subject to agreements; list of sole-source suppliers.
	4. Key software and technology licenses, by the company from third parties and vice versa.
	5. Overview of company's policies regarding trade secrets and proprietary information.
	6. Copies of material consulting agreements.
	7. Form of distributor/reseller agreements and list of all major distributors/resellers and related sales volume.
	8. Copies of joint-venture agreements.
	9. Copies of OEM agreements.
	10. Description of circumstances under which any material contract, agreements, or commitment is subject to renegotiation, termination, or default based on performance, change in control, etc.
	11. Details of any rights, consents, permissions, or licenses required by the company for carrying on its business.
	12. List and description of any government regulations that have special application to the business of the company.
	13. Details of all restrictions binding on the company that affect its ability freely to carry on its business in any part of the world.
IX. Insurance	
	A summary of all of the company's current insurance policies with details of renewal dates, premium, amount of cover, risks covered, name of insurance company, and of any outstanding or pending claim.
X. Litigation and Default	
	1. Details of any current litigation in which the company is involved (and that may affect the company) and of any that is pending or threatened.
	2. Details of any default under any material agreement, material trust deed, or other material arrangement.

X. Litigation and Default (cont.)	
	3. Details of any existing or pending judgment affecting the company.
	4. Details of any investigation or inquiry into or affecting the company by any government, administrative, or regulatory body.
XI. Defective Products	
	Details of all products manufactured and/or sold or licensed by the company that are known to be defective and details of all defective product claims for the last three years.
XII. Employees, Officers, and Directors	
	1. Agreements entered into over the last five years for loans to and any other agreements (including employment contracts) with employees, officers, affiliates, and directors including loans to purchase shares and consulting contracts.
	2. Any union contracts in effect and any correspondence relating to union organization efforts. Details of any dispute or industrial action by or with a union within the last three years.
	3. Copies of noncompetition, nondisclosure, and technology-assignment agreements between the company and its employees, officers, affiliates, and directors.
	4. Listing showing name, title, and compensation of those persons with current compensation over $100,000, together with summaries of all bonuses and other special compensation paid during the last two years.
	5. Overview of all executive and employee bonus plans, share-option plans, and profit-sharing plans.
	6. Form of sales commission plans and arrangements.
	7. List of claims in the last five years with respect to discrimination, unfair labor practices, or other employment-related factors.
XIII. Employee Benefit Plans	
	1. List of all ERISA Affiliates, such as one or more corporations or trades or businesses that, together with the company, are under common control within the meaning of Code Section 414(b) or (c); form an affiliated service group within the meaning of Code Section 414(m); or are entities whose plans must otherwise be aggregated under Code sections 414(n) or (o).
	2. List of retirement plans maintained by the company or any of its ERISA affiliates.

XIII. Employee Benefit Plans (cont.)	
	3. List of the trustee and plan administrator of each plan.
	4. List of nonqualified plans maintained by the company or any of its ERISA affiliates.
	5. List of welfare benefit plans maintained by the company or any of its ERISA affiliates.
	6. List of any welfare benefit plans covered through a cafeteria plan.
	7. Copies of the governing plan instruments for each plan (plan documents, group insurance policies, summary plan descriptions, etc.).
	8. Copies of the last three years of Form 5500s, including all schedules and attachments, as well as an independent auditor's reports and opinions.
	9. With respect to each qualified retirement plan (whether terminated or currently in existence), a copy of most recent IRS determination letter.
	10. List of all claims filed by a participant or beneficiary under any employee benefit plan, other than a claim for benefits in the ordinary course.
	11. Copy of any notice or similar communication from the Internal Revenue Service, the Department of Labor, or any other governmental agency relating to any employee benefit plan.
	12. Copy of audited accounts of any employee benefit plan audited by the Internal Revenue Service or the Department of Labor.
	13. Describe any development in regard to any employee benefit plan that could give rise to a significant risk of litigation or loss to the plan or the company.
	14. Copy of all employee benefit plans and programs and a list of compensation structure by employee grade.
XIV. Intellectual Property	
	1. Details of all patents, trade or service marks, registered design, copyright, or other know-how or intellectual property rights owned by the company or licensed or used by the company, and applications thereof.
	2. All development or other technology agreements to which the company is a party.

XIV. Intellectual Property (cont.)	
	3. Details of (and copies of documents relating to) all intellectual property in respect of which rights have been granted either to or by the company.
	4. Details of any complaints, objections, claims, or potential for alleged infringement or notified to the company or of which it is aware in relation to the ownership or licensing of intellectual property rights.
	5. Details of any service, facilities management, or outsourcing agreements relating to intellectual property used by the company.
	6. Details of any existing or threatened claims, actions, or proceedings brought by or against the company or any of its subsidiaries and relating to the use or possible use of any software, know-how, or intellectual property right or pending applications for them, including any proceedings that comprise or contemplate a challenge to the validity or grant of any intellectual property right.
	7. Details of any dealings by way of assignment, charge, mortgage, or other encumbrance, whether or not registered, in respect of any software, know-how, or intellectual property right or that has or may have effect to provide the company with less than full beneficial ownership of them.
	8. Details of all confidentiality agreements and agreements relating to the use of know-how to which the company is a party, whether as licensor or licensee or discloser or person to whom disclosure is made, and all agreements under which there is any restriction on the use or disclosure of know-how.
	9. Details of all third-party materials used by the company now or in the past for which no intellectual property right license has been obtained, specifying the reason in each case for not obtaining a license.
XV. Property	
	1. In respect of each of the properties owned or occupied by the company and/or its subsidiaries, please supply the following information:
	a. description;
	b. copy plan of the property edged in red;
	c. use;
	d. whether freehold or leasehold, and if leasehold:
	i. term;
	ii. rent;

XV. Property (cont.)	
	iii. permitted use;
	iv. rent review dates; and
	v. a copy of the lease and any head or subsidiary leases;
	e. whether title to the properties is registered or unregistered;
	f. whether or not the company occupies the whole of the property itself and, if not, please provide details of any sublettings and other occupational rights granted;
	g. mortgages/charges (including type of charge, name of mortgagee, the amount secured, and any conditions more onerous than those usually found in a charge executed by a corporate borrower over a commercial property); and
	h. a copy of any valuation obtained within the last three years.
	2. List of holders of title deeds to each of the properties and where they may they be inspected.
	3. Details of any notices (with copies) served by any tenant, reversioner, owner of adjoining property, local, or other authority.
	4. Details of all planning and zoning applications and permissions, by-law approvals, building regulations, listed building consents, and other notices and consents obtained from or required by any relevant authority relating to property owned or occupied by the company and any refusals received from such authority.
XVI. Environmental Matters	
	1. Purchaser will conduct a Phase I review.
	2. Copies of all Phase I or Phase II reviews done by or for the company in the past five years.

Editor and Author Biographies

About the Editors

Brian Culshaw was born in Lancashire, England, on Sept. 24, 1945. He graduated from University College London with a first-class honors degree in physics in 1966 and a PhD in electrical engineering in 1970. He has held appointments at Cornell University (1970), Bell Northern Research, Ottawa, Canada (1971-1973), University College London (1974-1983), Stanford University (1982), and has been professor of electronics at the University of Strathclyde since September 1983, where he set up the Optoelectronics Division within the Department of Electronic and Electrical Engineering and has acted as vice dean of the faculty of engineering.

Prof. Culshaw's personal research activities have centered upon optical fiber sensor system and network studies and novel technologies for optical fiber instrumentation, including pressure measurement systems, strain and temperature measuring systems, gyroscopes, and structural assessment techniques. He has also been involved in journal and book publishing in both editorial and author roles.

He was de facto technical chair of the first (1983) International Conference on Optical Fibre Sensors (OFS) and initiated European meetings in smart structures and the EWOFS workshop series in optical fiber sensor technology, the first of a series regarded as the definitive meeting. The two later consolidated in the corresponding communities. Predominantly with SPIE, he has organized numerous other conferences and workshops in Europe, the U.S., and Asia, including a precursor to Photonics Europe. However, undoubtedly the most challenging was OFS(C) in Wuhan, China, in 1991.

In 2004, Prof. Culshaw was elected secretary, then served as a director, and during 2007 as president of SPIE, the Society of Photo-Optical Engineers, the largest international society serving the interests of optical and photonics scientists and engineers. Based in the United States, in Bellingham on the Washington coast just south of Vancouver, BC, Canada, the society also has an office in Europe and extensive interests in the Far East.

Prof. Culshaw is a founding director of OptoSci Ltd., a spin-out that was established in 1994 with the primary objective of exploiting innovative optical technologies developed within University of Strathclyde. He has recently also co-founded Solus Sensors to exploit specific opportunities in the oil and gas sectors. He has consulted to numerous industrial organizations in the UK and abroad on fiber sensor technologies, their applications, smart structures, and related techniques. He has published in excess of 300 technical papers, several conference proceedings, seven textbooks, and a dozen patents.

José Miguel López-Higuera was born in February, 1954, in the village of Ramales de la Victoria, Cantabria, Spain. He obtained his technical engineering degree in telecommunications from the Universidad Laboral de Alcalá de Henares, Madrid and his telecommunications engineering degree from the Universidad Politécnica de Madrid (UPM). He obtained his PhD in telecommunications engineering, with an extraordinary award, at UPM. An assistant professor in the Universidad Laboral de Alcalá de Henares since 1976, and since 1986 at UPM, in 1991 he became an associate professor, and in 2001 a full professor in the University of Cantabria (UC), where he teaches courses in electronics and photonics.

He was the head of the Components and Technology Department, and director of the Technical School of Telecommunications Engineers in the Universidad Laboral de Alcalá de Henares. He was the director of the COIE and founded and heads the Photonics Engineering Group of the UC.

Prof. López-Higuera fabricated the first lithium niobate integrated optic devices in Spain and presently works in the development of new devices and subsystems for optical communications and sensors in photonic instrumentation, optical fiber sensor systems, and in materials detection and characterization using photonic techniques. He is currently manager and coordinator of R&D&I projects and has directed more than 40. He promoted the creation of the spin-off company TELNOS and was a member of the steering committee of the company Fibersensing.

He has also been involved in journal and book publishing in both editorial and authorial roles. He has written or co-written more than 400 publications in the form of books, chapters, papers, and conferences, and he has obtained several patents. He has acted as technical reviewer for several journals and conferences in the field of photonics.

He has been a member of the Technical Program Committee, International Steering Committee, technical co-chair, and general chair of scientific international meetings such as the International Optical Fibre Sensors Conference (OFS), the European Workshop on Optical Fibre Sensors (EWOFS), and the Optoelectronics Distance Measurements and Applications (ODIMAP), among others.

Prof. López-Higuera is a senior member of the Institute of Electrical and Electronic Engineers (IEEE), associate senior member of the Institute of Electrical Engineers (IEE), and member of the Optical Society of America (OSA) and SPIE.

About the Authors of Part I

Guillermo de la Dehesa was born in Madrid in 1941, and he studied Law and Economics at the Universidad Complutense of Madrid. Between 1978 and 1988, he served at the ministries of trade, industry, and energy and economy and finance, appointed successively as director general of trade, secretary general of industry and energy, secretary general of trade, secretary of state of economy and finance, and secretary of the Commission for Economic Affairs of the Council of Ministers. He also worked at the Bank of Spain as director for international affairs and foreign reserves management between 1980 and 1983. Since 1988, he has held different relevant positions in the private sector. At present, he is vice chairman of Goldman Sachs Europe Ltd, an independent director and member of the executive committee of Banco Santander as well as of Unión Fenosa and Campofrío in Madrid, and of AVIVA PLC in London. He is chairman of the Centre for Economic Policy Research (CEPR) in London; member of the Group of Thirty in Washington; chairman of the European Central Bank Observatory in Madrid; member of the Euro 50 Group in Brussels; chairman of the board of the Instituto de Empresa in Madrid; member of the advisory board of CREI at the University Pompeu Fabra in Barcelona; member of the scientific advisory board of the Instituto de Estudios Europeos and of the Instituto Elcano (both in Madrid); and chairman of the steering committee of the advisory board of ESCP-EAP in Paris. He is also a monetary expert to the economic and monetary committee of the European Parliament. He is has written nine books and coauthored another 31 books on economics. He has published more than 90 papers in economic journals and more than 300 articles in newspapers and magazines. He is a regular columnist at leading newspapers.

Stuart Barnes was born in Worksop, England. He attended Retford Grammar School, Nottinghamshire, and went on to read Engineering at Queen Mary College, London. On completing his PhD at the same institution, he joined Standard Telecommunication Laboratories (STL) in Harlow in 1978. He held a number of senior research posts at STL prior to becoming technical manager of STC Telecommunication Cable in Newport, Gwent. After that, he became deputy technical director of STC Submarine Systems (now a part of Nortel), based in Southampton, becoming technical director in 1993, where he was based at the Greenwich facility. Following the acquisition by Alcatel of the submarine business, he moved back into research, becoming deputy technical director of Alcatel Recherche based in Marcoussis, France. After three years in France, he returned to the UK and founded ilotron, a privately funded optical switching start-up based on the Essex University campus in Colchester. ilotron foundered in the post bubble meltdown, and he went on to become an entrepreneur in residence at Atlas Venture, a $5 billion-funded tier one venture capital fund. At Atlas, he cofounded Azea Networks in 2001 (where he is CTO) along with Scott White (previously of Atmosphere Networks), Steve Webb, and Dave Winterburn. Azea Networks has received about $50 million in venture funding to date and makes transmission equipment for upgrading submarine transmission links. He is a visiting professor of electrical engineering at Southampton University and supports the industrial advisory initiatives of the School of Electronics and Computer Science. He also works closely with the School of Electronic Engineering at Aston University. He lives in London and Salisbury with his long-suffering wife and three children. In his spare time, he plays golf badly and supports Southampton FC.

Mike Redman gained a BSc in Applied Physics on a sandwich course sponsored by the Ministry of Aviation. He then went to the physics department at Warwick University, where he was awarded a PhD for the study of radiation-induced color centers in various materials. After a short post-doctoral fellowship and an even shorter spell in the civil service, he joined Xerox Research (UK) as a technical specialist. He spent five years as a team leader in a group investigating novel printing and display technologies; this turned out to be a very productive time for filing patents.

After leaving Xerox, in 1979 Redman went on to become a founding member of the new-product-development unit of Standard Telephones & Cables. Here he worked on a number of new communications technologies, including display pagers, cordless telephones, voice messaging systems, and one of the first analogue mobile telephones. Rising to technical manager, he left in 1986 when the declining fortunes of STC brought about the unit's closure.

In 1986, he was recruited by Raychem to join the division that was developing Rayfos, a system for installing fiber-optic cables on electricity transmission lines. When Raychem sold this business to Cookson Group, Redman moved to the new company, Focas, which was set up to exploit the technology. He remained at Focas for nearly 10 years and was closely involved in many technical aspects of the company's business, helping it grow from a start-up to a company with a turnover of tens of millions of pounds.

In 1997, he was made redundant yet again and decided to start a company of his own, Ormal, to manufacture a range of sensor cables. His time at Ormal was not without its trials and tribulations, but the company is still very much in business, having evolved into Ormal Electronics and Engineering. Redman officially retired in 2005 and is now a consultant to Ormal and several of its customers.

Michael Lebby received his DEng, PhD, MBA, and BEng (Hons) from the University of Bradford, England, in 2004, 1987, 1985, and 1984, respectively. He received an honorary doctorate (DEng) from the University of Bradford in 2004 for his contributions to the field of optics, optoelectronics, and fiber-optic packaging. Since 1977, he has been employed by the British government, AT&T Bell Laboratories, Motorola, and AMP (now Tyco) in senior technical and business optoelectronic and fiber-optic roles. In 1999, Dr. Lebby joined Intel as a venture capitalist. In 2001, he founded a new fiber-optics company, Ignis Optics, where he served as CEO and board member. It was acquired by Bookham Technology in October 2003, where he was responsible for corporate and technical strategy. Dr. Lebby is now president and CEO of the OIDA (Optoelectronics Industry Development Association) based in Washington, DC. OIDA is known for its technology roadmaps, workshops, industry networking, and government policy. Dr. Lebby frequently provides tutorial lectures, speeches, and courses to senior executives on the dynamics of the optical networking environment. A Fellow of IEEE, Dr. Lebby also holds more than 170 issued U.S. patents and has published and presented regularly in the optoelectronics field.

About the Authors of Part II

Miguel Mulet Parada obtained an M.Eng. degree in Electrical and Electronic Engineering from Imperial College, London in 1996. In 2000, he obtained a D. Phil degree from the University of Oxford for his research on the automated analysis of echocardiographic data done at the Medical Vision Lab of the Engineering Science Department led by Prof. Mike Brady. Miguel was part of the original founding team of OMIA, which later became Mirada Solutions until its sale to CTI Molecular Imaging. At Mirada, he took part in technical and business development functions, lead the Cardiology project and managed the process to obtain market authorisation for the Fusion 7D product in the U.S. and Europe. He also has an MBA degree from the Said Business School at the University of Oxford and currently lives in Madrid, Spain.

Sir Michael Brady FRS, FREng is BP Professor of Information Engineering at the University of Oxford. Professor Brady's degrees are in mathematics (BSc and MSc from Manchester University, and PhD from the Australian National University). He was appointed Senior Research Scientist of the MIT Artificial Intelligence Laboratory in 1980 and helped found its world famous robotics laboratory. In 1985, he left MIT to take up a newly created Professorship in Information Engineering at

Oxford where he founded the Robotics Laboratory and the Medical Vision Laboratory (MVL). Prof. Brady combines his academic work, with a range of entrepreneurial activities, including non-executive board positions at Oxford Instruments plc and Isis Innovation (Oxford University's intellectual property company). Prof. Brady is a founding director of the start-up companies Guidance (http://www.gcsltd.co.uk), which develops navigation systems for mobile robots and for dynamic ship positioning as well as electronic tags for offenders, and Mirada Solutions Limited (http://www.mirada-solutions.com), which developed medical image analysis software and became Siemens Molecular Imaging in April 2005 after its purchase by CTI Molecular Imaging Inc. Most recently, Prof. Brady has been an investor and Senior Independent Director of the start-up companies Ixico, which provides image analysis services to the pharmaceutical industry, and an investor and chairman of Dexela, which is developing a novel 3D mammography system for more reliable and early detection of breast cancer. He is also consultant to the Translational Medicine and Genetics group of GSK, where he is an external coordinator of imaging in oncology.

Tom Baur is the Chairman of the Board for Meadowlark Optics Inc, which he founded in 1979. Meadowlark Optics designs and manufacturers a variety of polarization-based components and assemblies. These include; beam splitters, polarizers, retardation devices, liquid crystal based variable retarders, attenuators and polarization rotators. Meadowlark also manufactures a variety of spatial light modulators.

Jay Kumler started Coastal Optical Systems in 1991 with Marc Neer, George Fabich, and Wilhelm Geissler, and he has served as Coastal's president for 16 years. Additionally, he is president of the Liebmann Optical Company and serves on the board of directors of MEMS Optical and the American Precision Optical Manufacturer's Association. Mr. Kumler has published technical papers and articles on optical design, manufacturing and optical metrology, and currently serves on two advisory committees for SPIE. Jay lives in Florida with his wife Tammy and his two children, Joshua and Stephanie.

Colleen Fitzpatrick PhD is a Fellow of SPIE. In 1989, Colleen founded Rice Systems, which through the years was responsible for many innovations in high-resolution optical techniques. These included the first use of the photopolymer bacteriorhodopsin as an interferometric recording material and, with Lucent Technologies as a partner, the first integrated optical rotation sensor. She was also the key team member for the design of the environmental sensor system on Northrup Grumman's successful bid on NASA's Jupiter Icy Moons

Orbiter (JIMO) project. Since she retired in 2005, Colleen has been a partner of Yeiser & Associates, a high-technology consulting company in Huntington Beach, California. Colleen presently consults with the Armed Forces DNA Identification Laboratory on mitochondrial and Y-chromosome identification projects. Her favorite of these has been the identification of the unknown child whose body was recovered from the wreck of the Titanic. Colleen also consults with international investment companies on locating individuals and organizations with investment potential. Her books *Forensic Genealogy* and *DNA & Genealogy* have been featured on NPR's *Talk of the Nation* radio program and in *The Wall Street Journal*. She is co-owner of Rice Book Press.

Valentin P. Gapontsev PhD founded IPG in 1991. He has over 30 years of experience in the field of nonradiative energy transfer in rare earth ions and solid-state materials and is the author of more than 200 scientific articles and several international patents. Dr. Gapontsev holds a PhD degree in Physics from the Moscow Institute of Physics and Technology. He is a frequent invited speaker on the subject of fiber lasers. He is a member of the Optical Society of America (OSA), the Institute of Electrical and Electronics Engineers (IEEE), and the New York Academy of Science. In 1999, OSA honored him with its Engineering Excellence Award. He was a 2002 New England finalist for the Ernst & Young Entrepreneur of the Year and received an honorary doctorate of science from Worcester Polytechnic Institute in 2001.

James C. Wyant received a BS in physics in 1965 from Case Western Reserve University and his MS and PhD in optics from the University of Rochester in 1967 and 1968, respectively. He was an optical engineer with the Itek Corporation from 1968 to 1974. The then joined the faculty of the Optical Sciences Center at the University of Arizona, where he was an assistant professor from 1974 to 1976, an Associate Professor from 1976 to 1979, and a Professor from 1979 to present. In 1999, he became the Director of the Optical Sciences Center, and in 2005 he became the first Dean of the College of Optical Sciences. He was a founder of the WYKO Corporation and served as its president and Board Chairman from 1984 to 1997. He was a founder of 4D Technology and DMetrix.

Wyant is a Fellow of the Optical Society of America (OSA), SPIE, and the Optical Society of India. Wyant was the 1986 president of SPIE, and he has been an elected member of the OSA Board of Directors and Executive Committee. Wyant has received several awards, including the SPIE Gold Medal, the SPIE Technology Achievement Award, and the OSA Joseph Fraunhofer Award.

Mike Morris is president and cofounder of Ocean Optics, Inc., a leading supplier of solutions for optical sensing. Morris cofounded Ocean Optics in 1989 when he received $1.2 million in Small Business Innovation Research (SBIR) grants from the U.S. Department of Energy, and he has served as the company's president since its inception. A biochemist by training and entrepreneur by nature, Morris has nearly 30 years of experience in the design, manufacture, and marketing of technology-related products and services. Prior to founding Ocean Optics, Morris was the Associate Director of Technology Transfer for NASA's Southern Technology Applications Center, and he also worked as a sales representative for worldwide laboratory products supplier Fisher Scientific. In his tenure at Ocean Optics, Morris has participated in the development of hundreds of products, helped grow company sales to an annual rate of nearly $50 million, and assembled a workforce of nearly 200 employees at six locations worldwide. Today, Ocean Optics is recognized as the inventor of miniature fiber-optic spectroscopy and has sold more than 80,000 spectrometers since 1989. Morris has a BS in Marine Science from the University of South Florida and an MS degree in Biochemistry from Rutgers University.

Dr. Richard Claus is involved in the design and use of smart materials for advanced engineering applications. He currently works full-time at NanoSonic after leaving a chaired faculty position at Virginia Tech in 2006. During the 1970s, he was part of the first group at the NASA Langley Research Center to embed optical fiber sensors in advanced composites. In the early 1990s, he chaired the first International Conference on Smart Materials and led the sensor instrumentation effort on the DARPA Smart Wing program, a predecessor to the DARPA Morphing Aircraft effort. Since 1995, he has worked on new nanostructured versions of multifunctional materials, especially those for use as transducers in sensing systems. This work has involved the synthesis of nanoclusters and other molecules, the formation of thin-film and bulk materials from these molecules, and the investigation of structure/property relationships of the materials for transducer use.

Claus has received the ASME/AIAA Adaptive Structures Prize, the ASCE Norman Medal, the Charles Stark Draper Award, an SPIE Lifetime Achievement Award for work in smart materials and structures, and prizes from the Optical Society of America, the IEEE, and the IOP. He currently serves as editor-in-chief for the IOP journal *Smart Materials and Structures*, and he participates in the international Optical Fiber Sensor (OFS) conference.

Dr. Jennifer H. Lalli received her PhD in polymer chemistry in 2002 in the polymer synthesis, nanocomposites, and magnetics research group of Dr. Judy Riffle. She received her MS in polymer chemistry at Virginia Tech under Dr. Riffle in 1999 and her BS from Penn State in chemistry in 1996. The emphasis of her career was on the design, synthesis, and characterization of high-performance polysiloxane networks functionalized with controlled molar fractions of polar moieties as metal complexing polymers for thermally conductive adhesives. Prior to her education at Virginia Tech, she pursued her polymer science career in Avery Dennison's Chemical Division for two years. Her work at Avery Dennison focused on the design, synthesis, and characterization of emulsion polymerizations for pressure-sensitive adhesive materials. She also implemented the development and testing of PSAs for ISO 9001. Her career at NanoSonic has focused on the design and synthesis of specialty polymers and functionalized metal, ceramic, and magnetic nanoparticles to generate three-dimensional nanocomposites and self-assembled coatings for microelectronics packaging applications and thermal management. She is also responsible for the development of Metal Rubber, flexible conductive sensors for aerospace applications, and novel magnetic nanostructured materials for biomedical applications. At NanoSonic, she leads the development of Metal Rubber and business development with Lockheed Martin, Boeing, Avery, and IBM. Lalli has more than 30 journal and conference publications and four submitted patents, including two pending patents that cover Metal Rubber and its applications. This work was featured in the June 2004 issue of *The Economist* and August 2004 issue of *Popular Science*.

Jaspreet Singh formed the company Fiberonics. After completing his master of technology (MTech) degree in optoelectronics and optical communications from IIT Delhi in Dec 1989, Jaspreet Singh has worked in the field of photonics for more than 16 years. During the initial three years, he served as a senior scientific officer in the IIT Delhi Industrial R&D Division, in which his major contributions included developing photonics components based on side-polished single-mode optical fibers and lithium niobate waveguides, and the development of three software packages related to computer-aided characterization of optical fibers and components.

Ingolf Baumann is the founder and CEO of Advanced Optics Solutions GmbH, Germany. He established his company in 1998 as a licensed manufacturer for Bragg gratings and related components. Today, the company mainly focuses on components for telecommunications and solutions for optical sensing.

Before he founded his business, Baumann had worked as a researcher in the photonics group of the Dresden University of Technology since 1991. His research was mainly devoted to optical fibers, Bragg grating technology, and optical sensing. He earned his PhD degree in 2001 from the electrical engineering's faculty of Dresden University of Technology, with a dissertation titled "Design and realization of fiber-optic Add Drop Multiplexers based on Bragg gratings and fused couplers." He is the author and co-author of more than 20 relevant national and international publications in journals and conference proceedings, and he holds a couple of patents.

Eric Udd is president of Columbia Gorge Research (http://www.columbiagr.com), a company he founded in 2005 to help move fiber-optic sensor technology into fielded applications, where he has had 30 years of experience through education, consulting, and research and development. He made fundamental contributions to fiber rotation, acceleration, acoustic, pressure, vibration, temperature, humidity, and corrosion sensors at McDonnell Douglas from 1977 to 1993 and later at Blue Road Research, which he founded in 1993. The resulting technology has been deployed widely. Mr. Udd has edited two books for Wiley, *Fiber-optic sensors: An Introduction for Engineers and Scientists* (1991) and *Fiber Optic Smart Structures* (1995), chaired more than 30 international meetings, holds more than 40 U.S. patents, and has about 150 published papers. He has acted as principal investigator on more than 50 government and commercial fiber-optic sensor and communication contracts. Mr. Udd was elected as a McDonnell Douglas Fellow in 1987 and a Fellow of SPIE in 1989. He is a member of OSA, IEEE, and LEOS.

Daniele Inaudi received a degree in physics at the Swiss Federal Institute of Technology in Zurich (ETHZ). His graduation work centered on the theoretical and experimental study of the polarization state of the emission of external grating diode lasers, and he was awarded the ETHZ medal.

In 1997, he obtained his PhD in civil engineering at the Laboratory of Stress Analysis (IMAC) of the Swiss Federal Institute of Technology in Lausanne for his work on the development of a fiber optic deformation sensing system for civil engineering structural monitoring.

Daniele is cofounder and CTO of SMARTEC SA (http://www.smartec.ch) and CTO of Roctest Ltd. (http://www.roctest.com).

He is an active member of OSA, SPIE, IABSE, and fib, was chairman of the sensor conference at the annual SPIE International Symposium on Smart Structures and Materials, and a member of the organizing committee of the annual International Conference on Optical Fiber Sensors. Daniele is author of more than 100 papers, three book chapters, and editor of a book on optical nondestructive testing.

Nicoletta Casanova received her degree in Civil Engineering at the Swiss Federal Institute of Technology in Zurich (ETHZ) in 1994. From 1994 to 1999, she was active as Technical Director at the Laboratory for Material Testing, Istituto Meccanica dei Materiali SA (Grancia, Switzerland). From 1994 to 1996, she served as scientific co-worker at the Swiss Federal Institute of Technology in Lausanne for the development of a fiber-optic deformation sensing system for civil engineering structural monitoring. From 2000 to 2003, she was also director of the Laboratory Istituto Meccanica dei Materiali SA. She is currently the cofounder, director, and president/member of SMARTEC SA, a member of SIA (http://www.sia.ch) and SVIN (http://www.svin.ch), and the author of several papers and articles both in the scientific and management domains.

José Luís Santos was born in 1960 in Porto, Portugal. After graduating in applied physics (optics and electronics) in 1983, he received the PhD degree in the same field in 1993, both from the University of Porto, Portugal. Presently, he is the head of the physics department of the Faculty of Sciences of the University of Porto. Additionally, he has been the scientific coordinator of the INESC Porto Optoelectronics and Electronic Systems Unit since 1994, and the manager of INESC Porto Optoelectronics and Electronic Systems Unit since 1998. He is also the president of FiberSensing's shareholder board.

Alberto Maia was born in 1962 in Porto, Portugal. He graduated in electrical engineering from the engineering faculty of the University of Porto in 1985. He became a part of the INESC staff in 1986 as an R&D engineer. He was the manager of the hardware support team at INESC Porto between 1990 and 2000, and the group manager of the Electronics and Integration–Optoelectronic and Electronic Systems Unit at INESC Porto between 2001 and 2004. Presently, he is the COO and production director of FiberSensing.

Luís Ferreira was born in 1969 in Vale de Cambra, Portugal. He graduated in applied physics (optics and electronics) in 1991 and obtained the MSc degree (optoelectronics and lasers) in 1995, both from the University of Porto, Portugal. In 2000, he received his PhD degree in physics from the same university, after developing part of his research work in fiber-optic sensing at the physics department of the University of

North Carolina at Charlotte, U.S. Between 2000 and 2001, he was an invited assistant professor of the physics department of the Faculty of Sciences of the University of Porto. During 1999 and 2000, he was also an invited lecturer at the physics department of the University of Aveiro and at the physics section, University of Trás-os-Montes e Alto Douro. From 2001 to 2003, he was manager of the Advanced Development Unit at MultiWave Networks, Portugal. Presently, he is a senior researcher at INESC Porto and engineering director at FiberSensing. He is also a member of the scientific council of INESC Porto and a member of the remuneration committee of FiberSensing.

Francisco Araújo was born in 1971 in Mirandela, Portugal. After graduating in physics (optics and electronics) in 1993, he received the PhD degree in physics in 2000 (fiber Bragg gratings), both from the University of Porto, Portugal. Between 1999 and 2000, he was an invited lecturer at the physics department of the Faculty of Sciences of the University of Porto, where he served as invited assistant professor until 2001. Between 2001 and 2003, he was manager of the Fibre Optic Technology Unit at MultiWave Networks, Portugal. Presently, he is a senior researcher at INESC Porto and product development director at FiberSensing.

Pedro Alves was born in 1972 in Lisbon, Portugal. He graduated in 1995 in computers and electronics engineering–telecommunications at Instituto Superior Técnico, Lisbon. In 1997, he submitted his thesis for a master's in science in electronics and computers systems–electronic systems to the same university. He began his professional career as an applications engineer at National Instruments (U.S. and Brazil) between 1998 and 1999. He was the branch manager at National Instruments Portugal between 2000 and 2003 and regional sales manager at Alfautomazione between 2003 and 2004. A member of FiberSensing's board of directors since 2004, Pedro Antão Alves is the company's CEO and director of sales and marketing.

Anders Overgaard Bjarklev was born on July 2, 1961, in Roskilde, Denmark. He has an MSc in electrical engineering from the Technical University of Denmark (DTU) (thesis on optical communication, 1985), a PhD from DTU (thesis on optical fiber characterization, 1988), and doctor technices (DrTechn) from DTU (thesis on optical fiber amplifiers, 1995). Since 1985, he has been employed at the Technical University of Denmark, where in 1999 he was appointed Professor in Optical Waveguides at Research Center COM. In April 2001, he became a member of the Danish Academy of Technical Sciences, and in March 2004 he was appointed director of

COM-DTU, Technical University of Denmark. Anders serves as a referee on several international journals and has been supervising more than 30 PhD projects and more than 60 MSc thesis projects. He is author and co-author of two books (the latest published in September 2003) and two book chapters, more than 100 international journal articles, and more than 130 articles in international conference proceedings. Anders research interests are within the areas of dielectric optical waveguides, rare-earth-doped waveguide components, fiber amplifiers and laser sources, optical communication systems, planar waveguide structures, electromagnetic field theory, and photonic crystal waveguides. In 1999, he became a cofounder of the company Crystal Fibre A/S, and in 2003 he also became a cofounder of OCT Innovation Aps. He is a Fellow of the Optical Society of America.

Jes Broeng was born in Copenhagen on January 17, 1971. He received the MSc and PhD degrees from the Technical University of Denmark (DTU) in 1996 and 2000, respectively. His PhD work was focused on photonic crystal fibers, and highlights of his research include invention and experimental demonstration of the world's first optical waveguide operating by photonic bandgap effect. Jes received the annual award of the Danish Optical Society in 1999 and the European Optics Prize in 1999 for best scientific article in the field of optics. After finishing his PhD, Jes cofounded the company Crystal Fibre, where he today holds a position as manager of fiber research and development. Jes has co-authored more than 100 scientific publications and the first textbook on photonic crystal fibers.

Jose Salcedo PhD serves as CEO of Multiwave Photonics in Porto, Portugal. Dr. Salcedo established the company after 15 years of focused R&D efforts in optoelectronics, lasers, and fiber optics at the University of Porto, where he was professor of electrical engineering and previously associate professor of physics, and INESC Porto, which he co-founded in 1984 and where he also co-founded an optoelectronics center in 1990. Prior to this, Dr. Salcedo carried out professional activities in the U.S. for about 10 years, where he obtained MSc and PhD degrees at Stanford University in California.

About the Authors of Part III

Javier Ulecia graduated from the Universidad Politécnica, Madrid, as an aeronautical engineer. He holds an MBA from the H.E.C. School of Management, Paris. In the last 12 years, Mr. Ulecia has been involved in private equity and technology projects, either as an external consultant or as a manager of venture-capital funds.

In 2001, he cofounded Bullnet Capital, a venture-capital fund specializing in technology-related projects. Bullnet Capital is one of the few VCs existing today in Spain, and it has invested in five companies, including Netspira Networks (SW for mobile operators' data networks), Anafocus (vision systems), Arvirago Technologies (videogame developer), Multiwave Photonics (fiber optics), and GEM Imaging (nuclear medicine devices). In June 2005, Ericsson acquired Netspira Networks from its founders and investors, becoming one of the first examples of a Spanish company funded by a local VC that has been sold to a worldwide leader in its sector.

Javier is a member of the board of Bullnet Capital's portfolio companies as well as of other nontechnology companies. Before Bullnet Capital, he was CEO of Doing (a venture-capital firm that specialized in broadband technology operated until the technology bubble burst). Prior to that, he was senior manager at Bain & Company, the international strategy consulting firm where he worked first in the Paris office and, since September 1996, in Madrid. Javier began his career with Renault in France as a project manager in its Research and Development Division.

Joseph E. Gortych is president of Opticus IP Law, PLLC, an intellectual property (IP) law firm specializing in optics, photonics, and semiconductor technologies. He received his BS in physics from Rutgers University, his MS in optics from the University of Rochester's Institute of Optics, and his JD from Vermont Law School. Mr. Gortych is a member of the Vermont Bar and is a registered patent attorney. He is an active member of the Optical Society of America (OSA), SPIE, the American Bar Association (ABA), the Institute of Electrical and Electronics Engineers (IEEE), and the Licensing Executives Society (LES). Mr. Gortych has authored two books and numerous articles on the role of IP in high-tech business and continues to write, lecture, and advise on this subject in the U.S. and abroad. He is an inventor on six U.S. patents and on several other patents pending. Mr. Gortych has particular experience in assisting start-up enterprises with their IP matters as both inside and outside counsel. He also represents a number of larger, well-known optics, photonics, and semiconductor companies. His firm is based in Sarasota, Florida, U.S.

Chris Gracie, BSc MIET, graduated in 1966 with a degree in electronics and electrical engineering and joined Ferranti in Edinburgh, Scotland. The Ferranti company Gracie joined was bought in succession by GEC Marconi, BAE Systems, and Finmecannica. Gracie gained his wide experience of the optoelectronics industry during 30 years with the company. Following a career as a development engineer and program manager, Gracie became managing director of the company Displays Division, which designed and manufactured head-down displays, head-up displays, head-mounted displays, and video recording equipment, primarily for military aircraft. As Managing Director, he was responsible for the introduction of 12 new product ranges through the complete product-development cycle. Export was high on Gracie's agenda, and 17 new international markets covering Europe, Scandinavia, the U.S., India, and the Middle and Far East were developed in his tenure as MD.

Gracie left the industry in 1997 and became chief executive of the Scottish Optoelectronics Association (SOA). SOA was founded in October 1994 and has seen many changes to the industry and its funding mechanisms. SOA performs a number of services to its members, who currently number 90, in addition to operating projects under contract to Scottish, UK, and European government agencies. An important aspect of improving Scottish optoelectronics performance has been in strengthening the manufacturing process. Gracie is a director of OptoCap, a facility specializing in the development of packaging solutions for optoelectronics, microelectronics, nanotechnology, MEMS, biotech, micro displays, and sensors.

Gracie is extremely well connected in the electronics and optoelectronics communities worldwide in his roles as past chairman of the Institution of Engineering and Technology (IET), secretary of the UK Consortium of Photonics and Optics, and a founder/member of the International Optoelectronics Association (IOA).

About the Authors of Part IV

Prof. Gonzalo León received a PhD in telecommunications in 1982 and is full professor in the Telematics Engineering Department at the Universidad Politécnica de Madrid. His research activities were focused on software engineering for telecommunications systems and on technology transfer in the area of information society. Since 1986, he has occupied several relevant positions in the Spanish Administration of Science and Technology as deputy general director for international relations on R&D, where he was the Spanish delegate in CREST contributing to the definition of the R&D Framework Program and as deputy director at the Office of Science and Technology attached to the presidency of the government and responsible for the definition of the Spanish National R&D Plan (2000-2003). In 2002, he was appointed as Secretary General for Science Policy at the Ministry of Science and Technology, responsible for the National R&D Plan and international relations; he was responsible for the definition of the R&D Plan from 2004 to 2007. He has served as chairman of several high-level expert groups at the European Commission where, today, he is president of the Lisbon Strategy Group and vice president of the Space Advisory Group. Since 2004, he has been vice president for research at the Universidad Politécnica de Madrid.

Faramarz Farahi received his BS degree in physics from Aryamehr University of Technology in Tehran, his MS degree in applied mathematics and theoretical physics from Southampton University, and his PhD degree from the University of Kent at Canterbury, UK.

He is currently a professor of physics and optical science at the University of North Carolina at Charlotte, U.S., where he chairs the Department of Physics and Optical Science and is a member of the Center for Optoelectronics and Optical Communications. Dr. Farahi has more than 20 years of experience in the field of optical fiber sensors and devices.

His current research interests include integrated optical devices, hybrid integration of micro-optical systems, optical fiber sensors, and optical metrology. He also pursues research on the application of optical imaging and optical sensors in the medical field. He is the author or co-author of more than 150 scientific articles and texts, and he holds seven patents.